Gambling on Ore

Mining the American West

Duane A. Smith, Robert A. Trennert, and Liping Zhu, editors

Gambling on Ore

The Nature of

Metal Mining

in the

United States,

1860–1910

Kent A. Curtis

UNIVERSITY PRESS OF COLORADO
Boulder

© 2013 by Kent A. Curtis

Published by University Press of Colorado
5589 Arapahoe Avenue, Suite 206C
Boulder, Colorado 80303

 The University Press of Colorado is a proud member of
the Association of American University Presses.

The University Press of Colorado is a cooperative publishing enterprise supported, in part, by Adams State University, Colorado State University, Fort Lewis College, Metropolitan State University of Denver, Regis University, University of Colorado, University of Northern Colorado, Utah State University, and Western State Colorado University.

♾ This paper meets the requirements of the ANSI/NISO Z39.48-1992 (Permanence of Paper).

Library of Congress Cataloging-in-Publication Data

Curtis, Kent A.
  Gambling on ore : the nature of metal mining in the United States, 1860–1910 / Kent A. Curtis.
    pages cm — (Mining the American West)
  Includes bibliographical references and index.
  ISBN 978-1-60732-234-4 (hardback) — ISBN 978-1-60732-235-1 (ebook)
 1. Metallurgy—United States—History. 2. Metal trade—United States—History. 3. Mines and mineral resources—Social aspects—United States—History—19th century. 4. Mines and mineral resources—Social aspects—United States—History—20th century. 5. Ores—United States. I. Title.
  TN623.C87 2013
  338.2'74097309034—dc23
                                2013012681

Design by Daniel Pratt

22   21   20   19   18   17   16   15   14   13   10   9   8   7   6   5   4   3   2   1

Cover illustration of Pearl Street Station, circa 1883, courtesy, Smithsonian Institution.
Author photograph by Maxim Estevez-Curtis and Noah Estevez-Curtis.

For Marcela Estevez,
who made this book possible.

"The modern world is forged amidst our inattention."

Richard White, *The Organic Machine* (1996)

Contents

Figures

WHEN I FIRST BEGAN THIS PROJECT IN THE early 1990s, I thought I might have an interesting environmental history of industry to tell, explaining how the world's largest copper company ruined the wild places of western Montana. Two decades later, with a new global gold rush afoot and talk of mining the deep-sea trenches and deep-ocean mountain ranges once thought far beyond the miner's reach, I believe the importance of understanding the character of our social relationship with minerals is more relevant than ever. In that sense, this is an environmental history of mining. But it is not an anti-mining tome. Rather, it is an effort to examine mining as a critical relationship with nature, with particular attention to the character of mining that formed within the economic culture of the United States during the second half of the nineteenth century. I have come to believe that many of our modern ecological challenges result from the failure even of modern environmental reformers to acknowledge that ordinary everyday modern life, what we call "normal," rests entirely upon the social ecology of mining created during those years. Henry David Thoreau once wrote, "Our whole lives are startlingly moral. There is never an instant's truce between wrong and right." I don't think he meant this as an environmental statement, but I do think we can read an environmental lesson from it. Our lives and lifestyles are not separate from, but rather are wholly of, the natural world. We never stop communing with nature—no matter what we are doing, no matter who we are. The lesson I learned while researching this project is that we are all miners in the modern world, that mining made us possible.

Over the course of completing this story, the character of my own political commitment to environmentalism was transformed significantly. This was encouraged and enhanced by the powerful critique offered by postmodernism generally and Bruno Latour's work related to the sciences and democracy in particular. I began this work in the early 1990s with a sense of horror about the polluted landscape and a desperate desire to bear witness to the impacts and scars I found throughout the western mountains. I concluded, however, that the story of mining was most interesting in those elusive moments where miners or mining companies puzzled over the fundamental uncertainties of their work. I found that underneath the ecological wasteland left behind by the US metal mining industry, there existed a profound and intractable uncertainty.

My interest in copper mining began with the shock of a toxic landscape, massive fish kills, and miles of pollution but ended with the realization that the uncertainties of mining practices were ineluctable. So in the end, I did not write this book to convince people that we need to stop mining, which I once thought would be its purpose, but rather to help all of us see that we need to change our approach. As a fundamental environmental dependency, I hope this study makes clear the need to move beyond the improvised, unsustainable, and ecologically untenable pattern of ever-increasing metal output that has marked the US mining industry since the 1880s. In other words, I believe we need to mature as a mining society, not to reject our plight but instead to accept the profound responsibilities that come with it—something we begin to do perhaps by knowing ourselves somewhat better, by knowing ourselves as miners. At least that is my hope, and this study is my attempt at such an effort. I hope it causes readers to think about their place in the material context of the planet in ways they had not considered, and I hope it leads them to imagine new ways we all might work to improve these circumstances.

This book has traveled far since beginning as a news piece about the Clark Fork River in 1992. Too many people have been along for one part or another of this project for it to be possible to remember and acknowledge everyone. It has grown through two marriages and one divorce, nearly a half-dozen employers, and far more revisions and expectations than any project of this small scale should undergo. No one said history was an easy profession. History isn't easy. But the actual completion of the project does create a kind of indescribable euphoria that makes me want to at least try to recall many of those most critical to its final existence as a book.

In Montana—when a brash young college graduate decided to look into graduate school—Bruce Jenning, a political scientist and acting director of the Environmental Studies Program at the University of Montana, and Dan Flores, a rising young western historian with a terrific gift for story telling who was also at the University of Montana, provided invaluable support and encouragement at just the right moment and helped me make the terrifying leap from the comfortable space of journalism into the seemingly insurmountable challenge of graduate school and environmental history. They also helped open doors to a program at the University of Kansas that would change my life. In Kansas I had the privilege to study with the eminent historian and environmental scholar Donald Worster, whose keen eye, steady encouragement, and stellar example of hard work and great scholarship helped shape my ambitions for this work. Don was also a generous and inviting force of intellectual gravity in Lawrence, gathering faculty, post-docs, and graduate students into a vibrant orbit of conversation and rigorous academic pursuit. Among them, Paul Sutter, Mark Frederick, Jeff Crunk, Sterling Evans, Frank Zelko, Kevin Armitage, Mike French, Adam Rome, Amy Schwartz, Uwe Reisling, Maribel Novo, Roberto Irizarry, Nancy Scott-Jackson, Ben Dorfman, Sara Crawford, and Lisa Brady individually made Lawrence and my experiences there much less lonely.

Donald Worster not only influenced and attracted amazing graduate students to his program in Kansas, but his larger influence on the field of history—the idea that nature deserved a role as actant in our narratives—was articulated with such grace and conviction that early on the field attracted two important and influential historians, Richard White and William Cronon, whose graduate students were populating the American Society for Environmental Historian meetings in significant numbers by the early 1990s. The result was that throughout graduate school I was also given the gift of a pool of colleagues and budding young scholars around the country who were finding their own way through this emergent field and asking new and innovative questions. Among them, Matthew Klingle, Neil Maher, Jay Taylor, Sam Truett, Marsha Weisiger, Coll Thrush, Pamela Foster, Kathryn Morse, Nancy Langston, Thomas Andrews, and Fred Quivik contributed specific value to this project in ways they knew about and in ways they did not.

Throughout my research I had access to numerous archive collections of exceptional value, whose value was enhanced by the presence of thoughtful and encyclopedic archivists. None was more helpful than Peter Blodgett, H. Russell Smith Foundation curator, Western American Manuscripts, at the Henry E. Huntington Library in San Marino, California. In 2007 the Huntington Library awarded me the Mellon Research Fellowship, from which I gained in many ways and which made possible this final book manuscript, creating the bridge from dissertation to book. The library's manuscript collection included publications by mining engineers who had worked in or studied the Montana mines. The western history collection had every secondary source imaginable and dozens of rare books I had not known existed. The large fellowship program filled the library's reading rooms and lunch area, with dozens of scholars at all stages of the profession. Daily encounters were stimulating and enriching.

The book also benefited from two previous Huntington scholarships in 1999 and 2000, while I was developing the dissertation, as well as from grants from the Linda Hall Library of Science and Technology, University of Missouri in Kansas City, in 1999 and 2000. Having acquired the entire library of the Engineering and Mining Society, in effect, when it received Rossiter Raymond's professional collection in the 1990s, the Linda Hall Library gave me instant access to the textual world of nineteenth-century mining engineers. Bruce Bradley, director of the library, gave me free reign in that collection despite the fact that it was not fully cataloged, and he invited me to present my research at an open forum at the end of my stay. Thomas White, at the James J. Hill Library in St. Paul, Minnesota, provided a grant that not only allowed me to gather vital material about Montana's railroad development from that library's business collection but also gave me the time to dig into the Northern Pacific collection at the Minnesota Historical Society. Two unknown readers of this manuscript for the University Press of Colorado offered some of the most invaluable criticisms and encouragement I received during this entire process. This book is better because of their contributions and careful attention to detail.

No writer's life is complete without a mother like mine, Janet Beane, whose undying optimism and dogged confidence left me believing that something like writing a book was possible. It turns out, she was right. I owe a special debt of gratitude to my father, Robert Curtis, who insisted that I stay in school when I didn't want to and set me on the road of self-discovery that reaches a milestone with this book. He thinks I wasn't listening. I was given the gift of a large family and each of them—my sisters, Talitha Curtis Thurau, Johnanna Schimmel, Nicole Tracy, and Amity Beane and my brother, Zachary Beane—have walked with me somewhere along the way and carried some of the load writing this book entailed. I am grateful to have so many siblings and such generous people at that.

Several other individuals contributed to rounding out the manuscript over the past several years. Matthew Klingle, a fellow environmental historian and small-college professor (Bowdoin), who seems to knows every book ever written and every scholar who ever breathed history or philosophy from their lips and still has time to write his own award-winning scholarly works and read books in progress, has read more parts of this books in more drafts than he probably wants to remember. He has been there since this project first saw the light of day, and he has never failed to add value. Jacqueline Ayala, the most perfect undergraduate student a professor could want, undertook the laborious work of transforming Rossiter Raymond's detailed narrative in his mining reports into tables listing mining districts, mines, and ore production. Jacquie's hard work made it possible for me to see the patterns of change in the middle stages of lode mining in Montana. Sharon Hart and Ruth Pettis produced beautiful original graphics to help visualize some of the spatial and geological dimensions of the narrative.

In 2011 and 2012, Andrew Chittick and Erika Spohrer, treasured colleagues at Eckerd College, read early drafts of the final book and offered very helpful and trenchant commentary on a book outside their fields of expertise. Joel Thompson, a gifted scientist and generous colleague at Eckerd College, pointed me toward necessary geological studies to help me understand the nature of ore deposition. George Vrtis—fellow mining historian, fellow small-college (Carleton) professor, and a great friend whose encouragement and conversation over these past several years have kept my spirits up during the course of a long-term project—also read parts of early drafts and asked some great questions that helped me focus several parts of the rough manuscript. In the summer of 2011, Diane Gladu, office manager and fellow urban gardener, typed the final draft of the manuscript.

Many other people helped as well in many little ways, but I want to end by mentioning a few folks who really made it all worthwhile. My best friends Kenneth Resnik and Kenneth Tanzer, college roommates and lifelong buddies, saw this project through from its inception in the early 1990s until now, providing free couches, apartments, laughs, and at least a feigned interest in the intricacies of ore deposition. Everyone should be blessed with such friends. Their company meant more than I have ever acknowledged. And then there are the children, whom I have to thank. They are entering their second decade of life and have never known a daddy who wasn't writing a book about mining. Maxim Azucena and Noah Manuel made this project worth it by their very presence. I hope someday both of them can experience the overwhelming joy of having a son or a daughter or both, the reorienting, disarming, unbelievably grounding, unconditional love experienced in that context. They didn't know they were doing so, but they gave me that, and it helped me finish this book.

Most of all, I acknowledge and thank my life partner and soul mate, Marcela Estevez, mi amor, mi vida. Marcela carried the burdens of this project more than anyone, even more than I did at times. This book is dedicated to her for her unyielding love and support through all the years of gestation, for making this book possible. I am grateful beyond words.

Gambling on Ore

If we remove metals for the service of man, all methods
of protecting and sustaining health and more carefully
preserving the course of life are done away with. If there
were no metals, men would pass a horrible existence in
the midst of wild beasts; they would return to the acorns
and fruits and berries of the forest. They would feed upon
the herbs and roots which they plucked up with their
nails. They would dig out caves in which to lie down at
night and by day would rove in the woods and plains
at random like beasts, and as much as this condition is
utterly unworthy of humanity, with its splendid and
glorious natural endowment, will anyone be so foolish
or obstinate as not to allow that metals are necessary for
food and clothing, and that they tend to preserve life?

Georgius Agricola, *De Re Metallica*, translated from the first Latin edition
(1556) by Herbert Clark Hoover and Lou Henry Hoover (1950 [1912])

Mining, at least next to Agriculture, is of primitive and
essential interest to men, for it, alone, with the exception
of the arts of obtaining food, leads enterprise directly to
the supplies of nature.

Prof. F. H. Vinton, "Mining Engineering" (1874)

IN THE SPRING OF 1993 I stood in line at the University
of Montana cafeteria absentmindedly tapping my spoon on
the stainless steel countertop. I was working as an environmental reporter for the Missoula *Independent,* where I had
just filed a story about a massive fish kill in the nearby Clark
Fork River. The fish had been poisoned by arsenic that had
been scoured into solution from the sediments behind the
Mill Town Dam, a few miles upstream, when an ice floe had
made its way down the Blackfoot River and ground into
the reservoir sediments. The arsenic was residue from a century of copper mining and smelting that had taken place 120
miles upriver in Butte and Anaconda. I was puzzling over
the scope and scale of the environmental impact, admit-
tedly indignant at the arrogance of the copper interests and
dumbfounded by the seeming willingness of Montanans to
allow such impacts in their home landscape.

At the time, the Butte and Anaconda Superfund Complex—tens of thousands of acres of heavy-metal–laden earth, piles of mining waste, huge swathes of forest acid-burned to all but the most tenacious scrub, a half-mile-deep hole in the ground, and 120 miles of toxic river sediments—was among the nation's largest Superfund problems, a hazardous and toxic waste site of gargantuan proportions. I had learned that the first detection of arsenic in Mill Town's water had stumped the Missoula Health Department in 1981; the officials had no idea where it had come from. Some of them suggested a natural source, such as a vein of arsenic ore under the reservoir; others thought that perhaps a buried toxic dump, as had been found in the Love Canal community in upstate New York five years earlier, existed under Mill Town without anyone knowing. Only later were the upstream smelters identified as the source.[1]

The arsenic had come from the copper ores in Butte. Miners and geologists had long known that arsenic commonly existed as a natural by-product of copper smelting. In Butte, almost one-third of the copper ores were of a mineral called enargite, a compound containing half as much arsenic as copper. Smelting enargite had passed the arsenic out into waste dumps, up into the air, and down the Clark Fork River. As early as 1918 the main smelter in Anaconda was processing 65,000 tons of 6 percent copper ore a year, or about 1.3 million pounds of arsenic annually (a production level that continued to grow as ore quality continued to diminish throughout the twentieth century). In the most basic material sense, then, the quantities of arsenic were a factor of the quantities of copper produced. Given these known dangers, I wondered why so much copper ore would be processed in the first place. Many historians explained that production was necessitated by the invention of the telegraph, telephone, and electrical-generation systems, whose miles of copper wiring and various conduction needs required unprecedented amounts of the so-called red metal. The American and European desire for long-distance communication and domestic electrification created a demand for metallic copper to which Montana (and other western) copper developers had responded; the arsenic was an accidental by-product.[2]

But as I stood there tapping my spoon, I remained confused about how such a significant turn of events—the historically unprecedented excavation and processing of copper ores and the landscapes they produced—had been forgotten by late-twentieth-century Montana residents and had been somehow erased from the national stories told about Montana. The Montana I thought I had moved to in the early 1990s was a region celebrated for its wilderness areas—it contains some of the largest tracts of roadless land in the contiguous United States—and its national parks (Yellowstone and Glacier). Everyone I knew in Missoula celebrated the undeveloped wilderness characteristics Montana's mountain landscape provided, its whitewater rafting and backcountry skiing, its mile upon mile of untrammeled nature. This other, *industrial* history had been somehow masked in the region, except during moments like the fish kill, when it demanded a kind of sensationalized attention. "New Jersey with a view," was how my more cynical environmentalist friends described it before they jumped into their vehicles and headed for the nearest trailhead. I wondered why arsenic pollution and Superfund landscapes were not part of the general history of electrification and the rise of telephone communication or even, for that matter, the main part of the story of the rise of the mining industry in the United States during the late nineteenth century, as these stories seemed so obviously connected.[3]

While puzzling over these questions that day in the cafeteria, I had a flash of insight about them that would eventually lead to the present study. Mining had disappeared in plain sight. I stopped tapping my spoon and looked around at my surroundings: at the stainless steel countertop and serving utensils, at the eyelets in my shoes, at my belt, my glasses, the ovens and toasters and surfaces in the cafeteria, the window frames, wires, and cables; I thought about my car and my television and my personal computer. I realized for the first time, in a manner I had never considered before, that every part of my life depended to some degree on the refined products of mining. Metals were everywhere, and they were fundamentally necessary for the lives we live. It wasn't that I hadn't noticed metals before; it was rather that I hadn't considered them a part of nature.

My own ontological categories, the assumptions I made about how the world was composed, had led me to disassociate nature from the metals in my built environment. Up to that point, I had divided the material world into two large groups: natural things and artificial things. Natural things had their origins in nature and were produced by natural, organic processes. Artificial things, in contrast, originated in factories and laboratories and were produced by science, technology, and machines; they were artifacts of the human imagination and labor. This binary construction not only pitted the one against the other (the artificial *versus* the natural) but also had the cognitive effect of implying opposing forms of *origin*. Metals, as members of the category "artificial," would have no natural origins at all, or at least none with any substantial meaning.

My conceptual oversight became crystal-clear to me that day. I had operated with a limited understanding of the natural origins of metal, crediting its existence to science and technology and failing to imagine that the path of metals led back to nature. I had disassociated my own dependence on metals from the processes of ore formation and the practices of extraction, refinement, and fabrication into the artifice that surrounded and supported me. By seeing metals as artifice only, I had lost sight of their existence as a product of nature and of everything that natural existence implied. Montanans were not the only ones who had forgotten the devastating impacts of mining and smelting that made the ubiquity of metals possible; that ubiquity had been circumscribed by common sense itself. The modern categories by which most Americans had come to make sense of their complicated material world had created a material shortsightedness, an incomplete knowledge about the natural context of a minerals-based society. We have, oddly, forgotten that we are fundamentally a mining nation.

But for me, the boundaries that had separated the artificial from the natural suddenly began to appear less certain, less absolute. It became clear to me that day that metals are as natural as wheat flour, lumber, or sides of beef. I realized that our entire "artificial" infrastructure and the billions upon billions of tons of metallic materials that made such a world possible required and continued to require the relationship with nature that we call mining.[4]

Common sense suggested that I would uncover the deeper connections that wove natural minerals into modern society in the field of mining history, where there were hundreds of studies of western mining, including many that focused on Montana mining specifically. But this proved a bit of a dead end. Most of the vast scholarship in the field said very little about mining as an engagement and exploitation of *nature*; nor was much sustained attention given to the evolving processes by which the naturally existing enriched mineral deposits became the cultural commodity known as metal. In addition, these sources were often silent about describing these deposits. Instead, most of the studies constructed careful institutional, social, labor, and political histories similar to those done on any of several sectors of US business and political development. The studies collectively construct mining as something miners and mining companies simply did in places where large or valuable ore deposits existed. Mining seemed to emerge as a natural extension of national economic growth—although clearly, throughout all the studies, one can see an unusual degree of uncertainty and contingency in the practices.[5]

I turned next to the field of environmental history, a growing body of scholarship whose practitioners had set out to uncover the role of nature in US history. Here I found no studies of mining and only troubling guidance about how to frame the problem of mining within the discourse of environmental history. In the early 1990s, environmental historians had focused a large share of their attention on organic nature, studying agriculture, forests, parks, and the human habitat of cities, as well as the ideas and practices surrounding these dynamic ecological communities. Powered by the romantic-fueled wilderness ideal of nature, many of these studies constructed their stories against the background of a once-uncultivated, untouched landscape upon which they either traced a decline under the burden of the US economy or celebrated the preservation of some of these lands as a result of US conservation efforts. While these studies helped me understand the larger ideological reasons I had lost sight of the natural origins of metals, the approach threatened to reduce an environmental history of mining to a simple measurement of impacts. As a senior historian once asked when I told him I intended to write an environmental history of mining, "what else is there to say but that mining came in and tore things up?" Armed with the romantic wilderness ideal, an environmental history of mining could not be imagined as anything more than an environmental impact statement written backward. I didn't think such an approach would help us understand the nature of mining so much as make us horrified by mining's ecological outcomes.[6]

Serendipitously, however, at the same time I was beginning to frame these questions into a doctoral project, the field of environmental history was undergoing a process of rethinking and re-imagining that would ultimately enhance the kinds of questions I asked about mining as an environmental history. In particular, the historiographical concern with "nature" began to include questions of relationships, and narratives borrowed metaphors from physics and began to tell stories about the organization of space and the role of work in contributing to our ideas about nature (and not nature). At least one book-length study of mining followed immediately upon this turn, providing a wonderful critical analysis of the culture of the Klondike gold rush.[7]

Among these new studies, the most relevant to the questions I ultimately pursued was Richard White's *The Organic Machine*. In his short but provocative book, White reframed the question of nature in history as one in which relationships among energy, labor, and knowledge became the central focus. In my estimation, such a shift provided a workable alternative to studying environmental impacts for an environmental history of nineteenth-century metal mining. Rather than stories of abuse, White suggested looking for stories of relationship. Following Henri Lefebvre's very useful characterization of the living organism as an "apparatus which, by a variety of means, captures energies active in its vicinity," White had composed a succinct history of the Columbia River seeking out these *energy relationships*. His findings suggested that there have been historical configurations of knowledge, tools, landscapes, and ambitions, what he characterized as "energy regimes," whose interactions reveal the most intimate dimensions of the human-nature relationship. His narrative of the Columbia River uncovered not a widening gulf between people and nature, as the wilderness narrative has suggested, but instead a growing (if dysfunctional) intimacy, a hybrid tangle of human manipulation and natural processes, physical energies, and cultural narratives that can never be undone. White suggested that rather than characterize human use of nature as "rape," as so often happened in environmentalist discourse, "what had happened is closer to a failed marriage."[8]

Such a conception of environmental history gave my puzzle about Montana new life as a doctoral project and, eventually, as this book. I realized that my initial question about how people stood by and allowed their landscapes to be ruined was potentially misconceived and perhaps even distorted the relevant history of mining in Montana and, as I would soon discover, across the metals industry of the US West in the nineteenth century. I began to focus instead on the potentially more interesting and revealing stories about the processes by which Montana and the United States became so committed to mining in the first place. Clearly, the long-term environmental and human health outcomes that are metal mining's legacy today served no one, but like all failed marriages there had to be more to the story than the long-term outcomes. To understand the environmental history of mining from this perspective, it seemed prudent to explore western mining as the formation and development of a specific kind of social relationship with the natural world at a specific historical moment in a specific spatial and social context. *Gambling on Ore* is an attempt to do just that.

But writing about the formation and development of mineral exploitation as a social relationship in the nineteenth-century US West, it turns out, does not lead to an obviously "environmental" story. First, mineral deposits do not lend themselves to the familiar tropes and categories of environmental history. Few romantic poems have been written for ore lodes. Worse, the same mountains that are celebrated in the romantic tradition for their sublime presence, scale, and permanence are, through mining, excavated and even removed from the landscape; if ever an industry qualified for the "rapist" metaphor, mining would appear to do so. Further, the geological sciences have never been the traditional interpretative lens for environmental narratives. Rocks are, at best, the underlying and distant foundation creating the limits and possibilities atop which the real action of nature—ecology and community interaction—takes place; rocks have few stories of their own to add.[9]

In addition to the narrative challenges, the natural structure of mineral deposits—key actants, it turns out, in mining's environmental drama—makes them an elusive subject for both miners and environmental historians. Unlike forests and farms, cities and wildlife, mineral deposits are functionally invisible. For nineteenth-century miners, they were buried under solid, opaque earth. A miner could not have known whether valuable minerals existed in a particular location without prior investment of time, labor, and, as the nineteenth century wore on, money. In a delicious material irony, the only solid confirmation that a mineral deposit was valuable was the existence of valuable metal that was by then no longer a part of that deposit. When miners sought to profit from rocks, they entered into a kind of blind wager. For this reason, trying to synchronize the words and behaviors of miners with an actual physical mineral deposit is very difficult. Indeed, again and again in all forms of mining, one finds nothing but uncertainty at the interface of miners and rocks.

This persistent uncertainty made the mining relationship unstable; in the US mining industry after 1860, it created a recurrent set of patterns emerging out of mining's elusive successes. In short, US western miners developed an industry in which miners tended to over-invest in periods of uncertainty and to over-produce in periods of success. The one often led to the other, with some subset of the aggregate efforts yielding success and thus locating a site where overproduction would soon follow. Because these patterns were designed to exploit mineral deposits—hidden, specific, discrete, and, critically, finite geological material—overproduction meant the rapid exhaustion of a paying claim, which represented its own set of uncertainties and also contributed to industry dynamics. In the US West during the second half of the nineteenth century, these qualities engendered a growth dynamic whereby miners and mining companies continually pressed to intensify and expand production, well ahead of market demand. The overproduction of the age led many to characterize and understand the products of mining as the simple linear outcome of an extant natural resource endowment and made the presence of lots of metals something quite natural. Thus, in another rich irony related to mining, the almost unnatural overproduction of metals from US mineral resources led to the naturalization of metals in society. Because of the intensive pace of exploitation, the richest deposits disappeared very quickly and ore lodes either diminished in grade or disappeared, yet somehow the mining industry continued to wrest more and more metal every year from an ever-lower grade of ores.

Many of the stories that follow do not seem to have much to do with nature or environmentalist concerns; they are mostly about miners trying to profit from rocks. They divert waterways and fell forests and leave behind messes of tremendous proportion, but none of these impacts is my primary focus. Mineral deposits contribute to environmental stories by leaving a trace of their influence in the institutions and practices miners organize to exploit them. For this reason, *Gambling on Ore* focuses more on miners and mining institutions than on impacts and more on rocks and metals and water than on trees and animals. I have written a story that does not present itself as obviously "environmental" in the common sense of that term; that is, I am not narrating ecological decline. But in creating a story about what has become an inescapable relationship with nature in the modern world, I am writing what I believe is a very important environmental history.

Through this approach, I have come to believe that metal mining is not just another story to be told about the place and role of nature in US history; it also represents a keystone material relationship in the years that followed the US Civil War. In ways I hope will become clear in reading this book, as they became clear to me while writing it, mining established a set of approaches to natural resources that have come to define our production practices since that time. In this way, mining has had a profound influence on the human ecology and social relationships of modernizing North America throughout the twentieth century and the world after World War II. I believe that understanding how we forged these particular relationships is central to understanding the environmental history of the United States after 1850.

Gambling on Ore tells this story by studying the evolution of mining practices in the US West during the second half of the nineteenth century. Because my questions began in Montana, the main focus of this book is the Montana mining region, but it is offered as a representative case study of mining developments in the broader US mining west. Montana's mining history, like mining in the US West as a whole, can be divided into four major intersecting and overlapping mining episodes, each existing as a fundamentally different energy regime: the gold rush, the development of silver lode mining, the development of low-grade copper production, and the corporate consolidation of the regional base-metal production (in Montana's case, copper) industry.

In Montana, the accidents and uncertainty of the gold rush culture in the 1860s not only generated extreme acts of violence and chaotic, short-lived settlements; they also contributed to a shift to the extraction of silver ores and—after a period of adjustment, federal recognition, and the rise of a professional mining culture in the 1870s—began to express a pattern of iterative, uncertainty-related growth that came to mark the practice of hard-rock mining. The size and scale of silver lode mining by the 1880s not only brought deposits of copper ore into view; they also stimulated the belief that low-grade copper ores could be mined and processed for a profit in the western region. This belief did not generate immediate profits, but it did lead to some of the largest copper-processing facilities in the world, the enormous overproduction of metallic copper, and the early formation of modern business institutions in the US West.

But the modern copper industry could no more keep up with the perils of uncertainty embedded in the mining relationship than had any of the earlier stages of metal mining. Its efforts to control against these uncertainties only led to the production of landscape-scale impacts. The social conflict that followed as other users of the same landscape challenged the industry's right to diminish the fertility of their shared environment raised new levels of uncertainty for large-scale producers who had to justify their behavior in the courts. Only radical changes in the law and practices of adjudication that had, not incidentally, emerged in response to mining needs since the 1850s prevented the success of these social challenges to the kind of mining and smelting that had taken shape at the beginning of the twentieth century. By the time it became eminently obvious that mining had evolved into a set of destructive ecological relationships, the market products of these relationships had become too important for modern society to live without. This remains our conundrum.

The steps and stages on the way to twentieth-century, industrial-scale, mass-production mineral processing reveal a steadily growing commitment to an increasingly problematic undertaking, confronting proximate challenges only. At no individual point was it obvious how deep and destructive the mining relationship would become in the United States, but the anxious scramble to stay ahead of uncertainties and stabilize instabilities might have given us pause that something was afoot. Like failed marriages that can leave two people in utter despair, it was only when too much time had passed, too much water had flowed under the proverbial bridge, and too many wrong moves had been made that the pattern culminated in an obvious mistake. Unlike the metaphorical couple, however, the parties to this relationship were unable to go their separate ways; instead, as may be human nature in such circumstances, denial, deflection, and deceit worked to marginalize the ultimate results—except in moments when they insisted on reminding us they were still with us, like the arsenic in the Clark Fork River.

NOTES

1. "ARCO Environmental Action Plan for the Upper Clark Fork River Basin, Spring 1995," published by the Atlantic Richfield Company; Kevin Miller, "Arsenic Found in Milltown Water Supplies," the *Missoulian*, December 15, 1981; Miller, "Arsenic Probe Is Stepped Up, Warning Issued," the *Missoulian*, December 16, 1981; Miller, "Quantity and Toxicity of Arsenic in Milltown Top Previous Levels," the *Missoulian*, December 22, 1981; David Roach, "Contamination Leaves Residents Perplexed," the *Missoulian*, December 16, 1981.

2. Watson Davis, *The Story of Copper* (New York: Century Company, 1924), 37; G. L. Loughlin, *Mineral Resources of the United States: Part I—Metals, 1918* (Washington, DC: Government Printing Office, 1921), 270; Michael Malone, *The Battle for Butte: Mining and Politics on the Northern Frontier, 1864–1906* (Helena: Montana Historical Society Press, 1995), 34–35.

3. To give just two influential examples, neither Thomas P. Hughes, *Networks of Power: Electrification in Western Society, 1880–1930* (Baltimore: Johns Hopkins University Press, 1983), nor David E. Nye, *Electrifying America: Social Meanings of a New Technology* (Cambridge, MA: MIT Press, 1997), mentions copper ores or copper mining in their discussions of the rise of electrification in the western world.

4. Marx's labor theory of value makes a very similar point, although his concern was to recover the social activities (social labor) hidden behind the seemingly just-so market commodity, while mine is to recover the social relationships with *nature* embedded in these

hidden social activities. See Karl Marx, *Value, Price and Profit* (New York: International Company, 1969), esp. chapter 6, "Value and Labor." On the spatial expression of these three natural commodities, see William Cronon, *Nature's Metropolis: Chicago and the Great West* (New York: W. W. Norton, 1991), esp. part 2: "Nature to Market," 97–260.

5. There are many books about the California gold rush, but most of them either focus on the adventurous and sometimes paradoxical experiences of gold seekers leaving their homes and coming to new circumstances or treat the social, demographic, or cultural challenges confronting the communities that formed around gold mining in the absence of formal political structure and formal policy and regulation, but they rarely consider or even much describe the natural conditions of gold, the techniques enlisted for its exploitation, or the relationship between the two. See, for example, Mark A. Eifler, *Gold Rush Capitalists: Greed and Growth in Sacramento* (Albuquerque: University of New Mexico Press, 2002); May McNeer, *The California Gold Rush* (New York: Random House, 1994); JoAnn Levy, *They Saw the Elephant: Women and the California Gold Rush* (Norman: University of Oklahoma Press, 1992); Robert M. Senkewicz, *Vigilantes in Gold Rush California* (Stanford: Stanford University Press, 1985); Bernard J. Reid and Mary McDougall Gordon, *Overland to California with the Pioneer Line: The Gold Rush Diary of Bernard J. Reid* (Stanford: Stanford University Press, 1983); Ralph Mann, *After the Gold Rush: Society in Grass Valley and Nevada City, California, 1849–1870* (Stanford: Stanford University Press, 1982); J. S. Holliday and William Swain, *The World Rushed In: The California Gold Rush Experience* (New York: Simon and Schuster, 1981); Donald Dale Jackson, *Gold Dust: The California Gold Rush and the Forty-Niners* (Boston: Allen and Unwin, 1980); Robert E. Levinson, *The Jews in the California Gold Rush* (New York: Ktav, 1978); Rudolph M. Lapp, *Blacks in Gold Rush California* (New Haven, CT: Yale University Press, 1977); Andrew Bronin, *California Gold Rush, 1849* (New York: Viking, 1972); Rodman Wilson Paul, *California Gold* (Lincoln: University of Nebraska Press, 1965); Mary Floyd Williams, *History of the San Francisco Committee of Vigilance of 1851: A Study of Social Control on the California Frontier in the Days of the Gold Rush* (Berkeley: University of California Press, 1921). There are also a number of studies beyond California, stories about the discovery of ores in Nevada and the rise to fame of the silver barons on the Comstock Lode, such as future senators William Stewart and George Hearst, and stories about the millions made by Marcus Daly and future senator William Clark in the Butte copper fields. In all cases, the studies provide typical narratives of self-made men who learned how to best the uncertainties of mining and whose ambitions and character helped them join the ranks of the business and political elite during the aptly named Gilded Age, but these books also give few details about the natural state of the minerals exploited, and, other than profits and fortunes generated and sometimes measurements of the volume of metals produced, the social and environmental relationships comprising mining practice are all but neglected. See, for example, Richard Peterson, *The Bonanza Kings: The Social Origins and Business Behavior of Western Mining Entrepreneurs,*

1870–1900 (Lincoln: University of Nebraska Press, 1977); Russell R. Elliot, *Servant of Power: A Political Biography of William M. Stewart* (Reno: University of Nevada Press, 1983); Carl B. Glasscock, *The War of the Copper Kings: Builders of Butte and Wolves of Wall Street* (New York: Grosset and Dunlap, 1935); John Stewart, *Thomas F. Walsh: Progressive Businessman and Colorado Tycoon* (Boulder: University Press of Colorado, 2007). In the social histories of mining communities, mining serves as a backdrop within which settled working society organized itself against the harsh edges of industrial exploitation. This is also the case with other institutional and economic histories of mining companies that likewise present the act of mining, with the underlying environmental relationships as a given. See Mary Murphy, *Mining Cultures: Men, Women, and Leisure in Butte 1914–1941* (Urbana: University of Illinois Press, 1997); Philip J. Mellinger, *Race and Labor in Western Copper: The Fight for Equality, 1896–1918* (Tucson: University of Arizona Press, 1995); Susan Johnson, *Roaring Camp: The Social World of the California Gold Rush* (New York: W. W. Norton, 2000). Studies of mining labor show that mining was a specific and unique activity in nature, requiring specialized knowledge and understanding of the mining environment and the character of mineral deposits. As essentially labor histories, however, concerned with the formation and fate of the working class, these studies provide details of the underground workings only insofar as they help to reveal the consequences of deskilling as industrialization at the hands of finance capital gained control over the American economy. Thus they reveal little more about mining as a unique activity in nature or as a new US industry in the nineteenth century. As social histories of labor, these studies also show little interest in the broader ecological and geological forces influencing and being influenced by the rise of mining in the American West. See John Rowe, *The Hard Rock Men: Cornish Immigrants and the North American Mining Frontier* (New York: Barnes and Noble Books, 1974); David M. Emmons, *The Butte Irish: Class and Ethnicity in an American Mining Town, 1875–1925* (Urbana: University of Illinois Press, 1989); David Rosner and Gerald E. Markowitz, *Deadly Dust: Silicosis and the Politics of Occupational Disease in Twentieth Century America* (Princeton, NJ: Princeton University Press, 1991). Other exceptions include a very small handful of historical treatments of the broad pattern of mining development as it unfolded across the American West or across the United States generally. These studies reveal more of the kinds of large-scale changes that marked American mining development in the nineteenth century, tracing the growth of the industry from the first wave of the gold rush and placer mining in 1849, through the development of silver ore lodes in Nevada in the 1860s and elsewhere in the West in the 1870s, to the exploitation of complex copper ores in Montana, Utah, and Arizona in the 1880s and 1890s. These broader studies of the entire region help identify the geographic specificity of ore lodes and deposits and follow the steady import and development of mining skills and knowledge as mining district after mining district came under the productive energies of the American economy. They chart an interesting shift in the contours of American frontier development, as the slow but steady agrarian frontier that had pulled the United States westward through the first half of the nineteenth century gave way to a more rapid industrialized mining frontier that began in the Sierra Nevada foothills in California and leapfrogged its way eastward into the interior of the Rocky Mountains. As the industry came to dominate the western mountain landscape and to move into ever-more-complex ore lodes and lower-value metals, a parallel growth in the scale of mining operations and the engineering skills needed to develop the systems of extraction and processing embedded in these new mining systems tagged along, sometimes leading the way but more often following on the heels of expensive mistakes and waste of capital in less carefully planned investments. We finally learn about the environmental impacts associated with mining development every step of the way—from sand, gravel, and mercury pollution in California rivers, which impacted the budding agricultural interests of the Central Valley; to the tailings and other by-products of hard-rock silver mining; to the sulfur smoke and toxic pollution of the low-grade copper smelting choking entire communities in the valleys where mining took place. But from the earliest of these studies to the most recent, mining is framed as a story in which a given deposit of ore in the ground is unproblematically "discovered" and subsequently exploited by whatever mining interest is at hand. The mineral or ore was in the ground, as measured by its final production figures, and all the miners and mining companies needed to do was remove it as cheaply as possible and process it as efficiently as possible. The prior existence of minerals in the earth, in whatever volume they occurred, is presented as the unmitigated natural condition that drove the vast development of a mining industry in the US West in the nineteenth century. Everything else is narrated as a logical reaction to this "natural endowment," and the American public understood the impacts as the necessary results of this demanded exploitation, if they were thought to be understood at all. Interestingly as well, none of these studies makes any effort to link the mining and ore-processing activity to the broader material economy (to assess when and how demand emerged, for example). See Thomas A. Rickard, *A History of American Mining* (New York: McGraw-Hill, 1932); Rodman Wilson Paul, *Mining Frontiers of the Far West, 1848–1880* (Albuquerque: University of New Mexico Press, 1974); Clark C. Spence, *Mining Engineers and the American West: The Lace-Boot Brigade, 1849–1933* (New Haven, CT: Yale University Press, 1970); Duane A. Smith, *Mining America: The Industry and the Environment, 1800–1980* (Niwot: University Press of Colorado, 1993 [1987]). In fairness to the fields of mining history and environmental history, historians have already begun an effort to understand mining from this broader, more contextualized perspective. Two recent books suggest some of the contours and major themes that begin to arise with this kind of conceptualization of an environmental history of mining. The first is Andrew Isenberg's *Mining California: An Ecological History* (New York: Hill and Wang, 2005), which describes how patterns of industrialized nature in the California goldfields repeated themselves in gold mining experiences across the intermountain West and at the same time animated the region's timber and grazing interests with a similar spirit of industrialized natural resource exploitation. In other words, the techniques by which California gold rush miners solved the gold extraction problem in the Sierra Nevada foothills proved more broadly applicable to other natural resource exploitation efforts than previously acknowledged. Isenberg describes a "California example" in which large-scale technological control exploited resources as quickly as possible and usually with profligate waste, perpetuating a boom-and-bust economic development pattern that would define the American economy. "Mining" was both the original source of these impulses and an apt metaphor for the development spirit that followed. "Euroamericans reinscribed the political ecology of California upon the landscape of the West," Isenberg concluded (p. 178). In this way, the environmental history of western Montana's mining exists within a broader set of technical developments that began in and around California's gold creeks. Montana's gold rush took place nearly thirteen years after California's, but it recapitulated patterns of development very similar to those expressed in the Sierra Nevada foothills in the 1850s. Similarly, Timothy J. LeCain's more recent *Mass Destruction: The Men and Giant Mines That Wired America and Scarred the Planet* (New Brunswick, NJ: Rutgers University Press, 2009) not only offers a boldly conceived mining narrative that traces the cultural repercussions of brute force technology from its origins in open-pit mining out into the technological world of twentieth-century society, it also develops a story in which technological and ecological causes together produced a hybrid mining landscape. *Mass Destruction* follows the career of Daniel Jackling, who invented the techniques known as open-pit mining in response to increased scarcity of high-grade ores in US mineral fields. LeCain argues that overproduction of the richest ores in the West by the dawn of the twentieth century had created an industrial appetite for metals unprecedented in human history and soon to be starved by diminishing ore reserves. Jackling's technique filled the growing void between demand and supply, preventing scarcity and contributing to continued mining industry success and imagined new cultural ideas about power. LeCain's use of what is called an "envirotech" framework, an emergent perspective that seeks to merge the field of environmental history and the history of technology, contributed forcefully to his conception of mining technology as a particular cultural response to the conditions of nature as perceived by the industry.

6. Multiple studies explore the origins of environmental and ecological thought and policy, but some of the landmark studies in the field are Roderick Nash, *Wilderness and the American Mind* (New Haven, CT: Yale University Press, 1982); Samuel P. Hays, *Beauty, Health, and Permanence: Environmental Politics in the United States, 1955–1985* (New York: Cambridge University Press, 1987); Donald Worster, *Nature's Economy: A History of Ecological Ideas* (New York: Cambridge University Press, 1994); Paul Sutter, *Driven Wild: How the Fight against the Automobile Launched the Modern Wilderness Movement* (Seattle: University of Washington Press, 2002). See also Donald Worster, "Transformations of the Earth: Toward an Agroecological Perspective in History," *Journal of American History* 76, no. 4 (March 1990): 1087–1106. An example of the first kind of analysis can be found in

Donald Worster, *Dust Bowl: The Southern Plains in the 1930s* (New York: Oxford University Press, 1979) and an example of the second in Brian Donahue, *The Great Meadow: Farmers and the Land in Colonial Concord* (New Haven, CT: Yale University Press, 2004).

7. Kathryn Morse, *The Nature of Gold: An Environmental History of the Klondike Gold Rush* (Seattle: University of Washington Press, 2003). William Cronon, "Kennecott Journey: The Paths out of Town," in William Cronon, George Miles, and Jay Gitlin, eds, *Under an Open Sky: Rethinking America's Western Past* (New York: W. W. Norton, 1992), 28–51, also examined mining through the lens of environmental history.

8. See also Henri Levebvre, *The Production of Space,* trans. Donald Nicholson-Smith (Cambridge, MA: Blackwell, 1991 [1974]), 176; Richard White, *The Organic Machine: The Remaking of the Columbia River* (New York: Hill and Wang, 1995), 29, 59.

9. See, for example, "Prologue: Rocks and History," in Ted Steinberg, *Down to Earth: Nature's Role in American History* (New York: Oxford University Press, 2008), 2–7.

The humblest observer who goes to the mines sees and
says that gold-digging is of the character of a lottery.

Henry David Thoreau, "Life without Principle" (1863)

We deprecate the unreasonable excitements in the throes
of which new [mining] districts are brought forth, but
we confess that we should be puzzled to find a substitute
for them.

Rossiter Raymond, *Engineering and Mining Journal* (1874)

In Montana, as elsewhere, the placer miner blazed
the trail for the lode miner; indeed, widening the
generalization, one may say that all over our western
domain, as in other parts of the world, the finding of gold
has been the first step in the development of a mining
industry.

Thomas Rickard, *A History of American Mining* (1932)

ONE

*Producing a Mining
Landscape*

*Gold Rush Uncertainty
in Proto-Montana*

GOLD RUSHES MARK THE ICONIC beginning of mining stories in the nineteenth-century US West. In them,
gold seekers set out into uncharted territory seeking
untold riches said to be buried in mountain creek beds;
they chased rumors and stampeded; when a gold deposit
was uncovered, they excavated it with incredible speed.
Placer gold mining, as their activity was called, required
no advanced technology; according to historian Malcolm
Rohrbough, tools and techniques were widely shared
among miners. "Unlike many other economic enterprises,
there was little suggestion that superior techniques produced better results," he wrote. "Almost every miner in
the diggings did the same thing." They dug holes, channeled water, and washed dirt, hoping to find gold in their
pans, rockers, and sluices at the end of the washing. Gold
seekers also created and replicated a social technology, an
institution called the *mining district*, by which method
gold seekers organized themselves into a collective quasi-
political unit and the creek beds into regular placer claims.
When a district filled up or a creek was mined of its gold,

DOI: 10.5876/9781607322351.c01

miners moved on to other locations, prospecting mountain streams until they found gold once again and formed a new district. Through these processes, between 1849 and 1860, which Rodman Paul has rightly called the mining frontier, the mineral west came into being. By 1860, gold seekers had started and abandoned small and medium-sized gold mining settlements in California, Nevada, Colorado, Oregon, Washington, Idaho, and New Mexico.[1]

Montana's gold rush began while the eastern slope of the northern Rocky Mountains was still a part of Idaho Territory, and it happened by accident. Gold seekers traveling west through the region in the spring of 1862 heard rumors of gold deposits in Gold Creek in the northern Deer Lodge Valley, where cattle ranchers had fattened cows for almost a decade. A small preliminary group of successful miners attracted others to the location, setting in motion a gold rush that would literally bring Montana into existence. Beginning with a mere forty-five miners crowded into a small placer creek in June 1862, the discovery phase of the Montana gold rush unfolded over three years with the sequential opening of extensive gold deposits in three major locations: Grasshopper Creek, where Bannack City was founded in 1862; Alder Gulch, where Virginia City was founded in 1863; and Last Chance Gulch, where Helena was founded in 1864. The Territory of Montana was created in 1864, and by 1865 Montana Territory held more than 30,000 new settlers, almost all of them mining gold or somehow serving gold miners.

Underneath the broad strokes of discovery and exploitation, however, lurked a much more contingent story of gold deposits, gold seekers, and a string of accidents and violence that animated these years. Montana's gold rush experiences provide critical insights into the evolving social relationship with minerals that formed the foundation of the mining west. As one of the nation's preeminent mineral regions, Montana's story is offered as a detailed case study of the meaning of this unique nineteenth-century phenomenon, where men went to work to wash money out of holes in the ground.

Mining historians have traditionally looked at gold rushes as anomalous phenomena, moments of discovery followed by ephemeral social reactions. To the extent that environmental historians have considered gold rushes, they have deconstructed the cultural meaning of gold as value, noted the physical investment of labor by men in the gold creeks, and marveled disheartened by the scope and scale of the environmental impacts of placer gold mining. Another line of environmental inquiry, the one taken here, is to start with gold in nature to understand how it would have been encountered by miners. To do so reveals an uncertain geography for prospectors and exposes the random and invisible quality of placer gold deposits. Driven by *the idea of* gold but ultimately constrained by the physical realities of gold's deposition in the landscape, gold seekers and the societies they built reflected the deeply uncertain and unstable relationship with nature resting at the core of every gold rush and at the genesis of the US metal mining industry.[2]

THE NATURE OF PLACER GOLD

Like everything on earth, gold has a history. The gold placer miners would exploit in the US West after 1850 came from quartz deposits that had formed during the mountain-building episodes, or orogenies, out of which the western mountains had emerged beginning 170 million years ago. These quartz deposits had resulted from the geodynamics of plate tectonics and the play of atomic energies deep within the earth. Orogenies occur where and when two or more continental plates collide, causing buckling and folding, doubling over in places, cracking, lifting, and falling in others along the zone of intersection. I can describe an orogeny with these familiar verbs (to buckle, to fold, to lift, to crack), but the actual scale and pace of change is difficult to imagine. Orogenies take place over periods of 10 million to 20 million years, ever so slowly and imperceptibly lifting massive slabs of rock, folding others underneath, and gradually moving gargantuan chunks of continent.

The quartz deposits precipitate as the mountains form. Some large pieces of continent angle and dip beneath others and are forced into the asthenosphere, the layer of hot, viscous material beneath the earth's outer rocky crust. The depths bring both heat and pressure, whose combination liquefies the continental rocks like a giant smelting furnace, creating masses of molten continent, or magma. Hydrothermal solutions of silicon and trace metals precipitate out of these masses toward the surface, squeezing into whatever spaces are available between the top of the magma and the overlying crust. As the crustal slabs scrape and shear against one another, fissures open in the fracturing rock above; when they do, the hydrothermal solution expands, filling the openings in the rock. The expansion lowers the pressure in the material, cooling it somewhat and causing the quartz and its constituents to begin to precipitate out, eventually filling the cavities with a mineral deposit known as a *quartz lode*. Mountain formations are laced with these deposits, and the western mountains were no different.

Quartz deposits were everywhere in the mountains, but quartz deposits containing gold were not. In fact, only a few rare deposits in difficult-to-determine places carried traces of the metal. In addition, quartz's association with gold followed no particular or predictable pattern throughout the mountains. Some orogenic quartz deposits were enriched, but most were not.[3]

The placer gold deposits themselves—thin layers of gold dust and gold flake hydrologically sorted in the bottoms of creek beds—are tertiary deposits. They are the re-deposition of enriched quartz lodes through atmospheric erosion and water drainage. In the Rocky Mountains over 170 million years, ice and snow had cracked rock mechanically, acid rain had eaten at it chemically, and melting, running, falling water had conveyed these pieces down the mountains. As the waters washed the eroding mountains into canyons, forming creek beds out of the detritus and shaping waterways through the lowest points of the slope, the eroded material became sorted by size and weight, with the smallest and heaviest materials collected at the bottom. In the rare places where gold had been contained in the quartz, this sorting action would lead to the concentration of small particles of gold flakes and dust into a muddy, sandy layer deep beneath the creek detritus atop the hard pan basin in which the creek had originally formed. These were the so-called placer deposits that would be exploited in the series of gold rushes that opened the mountain west to exploitation and influenced the character of settlement in the US West during the thirty years after 1850.[4]

According to economic geologists, mountains on earth erode completely over a period of about 600 million years. In the United States, the Appalachian Mountain chain in the East is roughly 300 million years old, having once stood taller than the present-day Himalayan Mountains and possessing small but generally insignificant gold deposits in historic times. Most of their gold deposits had eroded into the sea long before humans came to the Western Hemisphere. The mountain ranges of the US West are much younger. They have experienced significant erosion, but they are not close to disappearing. The Rocky Mountains began forming roughly 170 million years ago with a series of sequential, overlapping mountain-building episodes beginning with the Nevadan Orogeny and ending with the Laramide Orogeny, which came to a close roughly 60 million years ago—shortly after the end of the Jurassic period and the extinction of the dinosaurs. In the western mountains, then, the timing was geologically perfect for the existence of concentrated tertiary gold deposits in the mountain creeks.[5]

Gold seekers sought metallic gold in the nineteenth century, as we understand, because it represented a natural form of money: it could be used to buy things, and it symbolized wealth. But the nineteenth-century Americans, Europeans, Chinese, and others who sought this material did not invent this cultural role for gold. Indeed, gold's rarity and physical qualities seem to have contributed to its cultural value in nearly all of the human societies that have encountered it for almost 6,000 years. Gold is extremely rare. Geologically, it comprises perhaps one millionth of 1 percent of all of the material in the earth's rocky crust, and its distribution throughout the crust is neither even nor regular. Most of the crust does not contain any gold. Where it is found, however, it has indelible physical qualities that contribute to its social value. Gold is a durable, pliable, and incredibly stable metallic element that resists both rust and corrosion. Its physical durability and chemically non-reactive

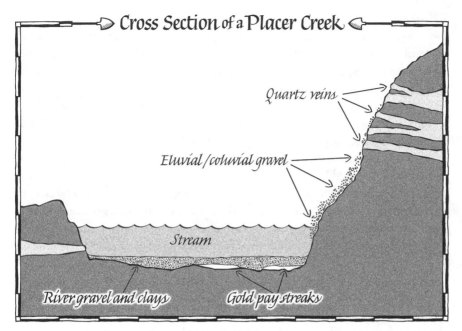

Cross Section of a Placer Creek

Quartz veins

Eluvial/coluvial gravel

Stream

River gravel and clays

Gold pay streaks

Cross-section of a placer creek. The gold sought by placer miners in the western US mountains rested in random locations at the bottom of creek detritus. Original artwork by Ruth Pettis.

qualities mean simply that gold retains its shiny physical form permanently, and it seems reasonable to presume that gold became a medium of exchange because of these qualities. Indeed, it is tempting to postulate that its quality of permanence could have contributed to the cultural concepts of abstract value in the first place. While Kathryn Morse is correct to argue that "the nature of gold, its bundle of physical characteristics[,] did not determine how human beings would use it," it seems equally clear that gold's physical characteristics, un-tarnishable surface, and physical stability lent material credence to a belief in the existence of objectively permanent economic value. In any event, by the nineteenth century gold's inherent value was an unquestioned social reality.[6]

But it is a mistake to imagine that a simple cause-and-effect relationship existed between placer gold mining and enriched placer deposits in nature. The use of the noun *discovery* to describe the causes of a gold rush leads to that misconception by imagining the gold in place, positing the knowledge of gold's presence, and then following the race to get to it first. It seems logical, even necessary, that members of the culture that occupied the places where gold was deposited would have scrambled to exploit it. But this tells the story backward. The actual gold deposits eventually unearthed across the mountain west had no initial social existence. We can describe their formation now in hindsight. Ironically, we know where and how much placer gold was excavated *because of* the very acts of mining purported to have been caused by it. Placer mining did not begin with the knowledge that a creek was rich in gold; rather, it produced that knowledge.

For this reason, then, gold rushes proceeded along the lines of *un*certainty. Placer mining involved a lot of guesswork and poking around. Even in the few places where gold had been deposited in significant quantities to warrant excavation, there were few, if any, outward signs of its presence. In the California Sierra Nevada foothills—the oldest gold-bearing mountains in the US West—the original quartz lodes had been entirely eroded into the creek placer deposits, and there was nothing particularly suspicious about the creeks themselves. In subsequent placer locations in Nevada, Colorado, Idaho, and Montana, some remnants of quartz deposits remained in some parts of the hillsides, but all that miners could see were weathered outcroppings of quartz that provided few clues in and of themselves and which appeared like thousands of other weathered quartz outcroppings that contained no gold. Indeed, in almost all cases, these outcroppings would not become obvious to miners until *after* the placer creek deposits had begun to be uncovered. Even in places where gold had begun to be washed out of the dirt dug out of creek beds, its random deposition pattern made it as likely as not that more of the same material lay nearby. Any secure knowledge of gold's place in western creek beds was by necessity retrospective and historic. In the end, it was impossible to know precisely where gold had been deposited and how much gold was actually present without doing the work of digging it out and washing it free of detritus.

But the work mining required could not take place in the US West without two important preconditions: the construction of transportation routes to the mining region, and the creation of a social and political space within which gold seekers could perform their work without fear of theft, sabotage, or murder. With the US victory over Mexico in 1848 and the bloody suppression of California Indian groups, the stage was set for the California gold rush, a defining moment in US environmental history. Similar forces left western Nevada similarly available in the 1850s. By 1858 the Northern Cheyenne, who had quickly chased the first gold seekers out of Colorado in 1849, had ceased to be a threat along the eastern front of the Rocky Mountains, creating conditions for gold exploitation in that region. With the end of the Columbia Basin Wars and the subsequent construction of the Mullan Road through the northern Rocky Mountains and federally subsidized steamboat traffic up the Missouri River, the northern Rocky Mountains became available after 1860. In each case, mining rolled into regions made accessible through state production of gold mining space.[7]

The production of a gold mining frontier, however, only marked the beginning of the *possibility* that a gold rush could safely take place; the work of mining still lay ahead and continued to be riddled with uncertainties. Gold seekers were uncertain about the location of gold deposits and, when gold deposits were found, they were uncertain about the contents of their claims. They would work incessantly and at times unsuccessfully for answers to both questions. The groups of men who formed teams of placer miners in these regions were also uncertain about the character of the new society that had formed around gold extraction. The combination of these uncertainties often generated extreme responses.

Built atop the product of orogenies, states, and cultural imagination, the western gold rush period worked against uncertainties and, in the process, both mapped and shaped the ethos of an entire region and eventually of the nation. All of the work placer miners did, in turn, left more of a legacy than previously acknowledged in both mining and environmental history. To make this case and expose the forces of influence that trickled up and spread out of placer gold mining activities in the US West, the remainder of this chapter looks carefully and with some detail at the individual and social experiences that comprised Montana's gold rush.[8]

THE STUART BROTHERS

James and Granville Stuart, brothers from Iowa, had joined the gold rush to California from the east in the early 1850s. After surviving for a few years in California as cattle traders, they had started a return trip home in 1857 when they were pushed north into the northern Rocky Mountains by the Mormon Wars. By 1860 they had established American Fork, a tiny settlement of two cabins and an outbuilding in the northern reaches of the Deer Lodge Valley, where they traded horse with cattlemen, mountain men, soldiers, and Indian groups. They knew from rumors and from their own prospecting work—as had most people in the region since at least the early 1850s—that the region's creeks held gold, but the difficulty of access, both in terms of sheer distances and the rugged topography of a mountainous region, as well as the real threat of violence (since most of the region remained sovereign Native American land until 1859), had kept miners at bay.[9]

But by 1860 national forces were transforming the northern Rocky Mountains from within. In 1859, after a swift and severe military campaign against a small coalition of Columbia Plateau Native American groups, the US Congress had ratified a set of treaties that opened the northern Rocky Mountain landscape to legal occupation by US citizens. Congress at the same time appropriated funds to support the engineering of a wagon road surveyed by Captain John Mullan a few years earlier. As road construction began, gold seekers flooded from the west into creeks feeding the Clearwater River along the western slope of the northern Rocky Mountains in present-day Idaho, and by 1860 several of the gold seekers had begun to produce gold dust from several locations. By the fall of 1861, men began trickling east from these creek diggings with caches of gold. In response, Congress helped one more time, offering contracts to private steamboat companies to haul mining supplies and potential gold miners up the Missouri River the following summer, when the river was in full flood, to Fort Benton—the highest point of steamboat navigation on the Missouri and the most distant water port in the United States.[10]

During the winter of 1862, a few months before the rush to the region began, the area now contained in western Montana had an estimated population of 12,430 people, 12,000 of whom were members of one of the several Native American groups that had occupied the region for decades and in some cases centuries. Most of the Natives were Blackfoot, with an estimated population of 10,000, but about 2,000 other indigenous people were divided among the Nez Perce, Flathead, Kutenai, Snake, Bannock, and Pend d'Orielle. The remaining settled population of about 100 people was made up of trappers, mountaineers, traders, missionaries, and a few recent settlers with Indian wives and mixed-race children—like the Stuart brothers and the Grant family, who owned a huge cattle herd that grazed on the rich grasses of the Deer Lodge Valley. To this was added the anomalous concentration of 330 soldiers, engineers, and laborers making their final winter camp in Hellgate, near present-day Missoula, after putting the finishing touches on the wagon road they had built through the northern mountains.[11]

In the spring of 1862 a group of six speculators had arrived in American Fork, at the time named Benetsee Creek after the French-Indian miner who had first discovered gold there in 1852. They hired Granville and James Stuart to cut and saw 1,000 feet of board. The creek above American Fork had been prospected the two previous summers (1860 and 1861) by Gold Tom, one of the six speculators, and he had successfully worked quite a bit of pay dirt in a crude sluice fastened together with wooden nails. As soon as the Stuarts completed the boards, the speculators went to work excavating and washing dirt in their sluices and prospecting additional ground.

When Captain John Mullan and his road team arrived along the new wagon road in late May, the new miners were enjoying steady success. "We halted at the American Fork and visited the Deer Lodge gold mines," Mullan reported, "where we found Messrs. Blake, McAdow, Higgins, Dr. Atkinson, and Gold Tom at work sluicing, and at that time they were taking out about ten dollars per day to the hand, and with fair prospects of extensive digging. Wherever parties had prospected in ravines or river bottoms they had found prospects from one to fifteen cents to the pan." This was important knowledge for Mullan to secure. It was one thing to hear a rumor of gold or to see men in possession of gold they claimed to have mined. It was quite another thing to see the gold itself being produced from creek detritus. Mullan reported: "Convincing ourselves by indubitable proofs of the existence of gold at this point, we continued our journey."[12]

Within a month of Mullan's visit, gold seekers had begun arriving at American Fork. On June 24 a large group of men arrived by horseback and pack train from Colorado, having been told of the Deer Lodge gold diggings by Thomas Stuart, Granville and James's brother, to whom they had sent a letter about Gold Tom's work the previous fall. "There are about forty-five emigrants in the mines at this time," Granville wrote on July 5. On July 11 James Stuart noted continued population growth: "Many emigrants arriving every day now." Two days later the miners renamed Benetsee Creek "Gold Creek," organized a mining district, and held an election in which James Stuart became sheriff. In late July the Gold Creek miners' court sent James Stuart after a horse thief, whom he captured and brought back to town to stand trial. The miners at the next Sunday's meeting took pity on the young man, whose defense was that he was down on his luck, and sent him out of town after collecting fifteen dollars in donations to help him along.[13]

As men continued to arrive at the creek and more often than not encountered initial failure, the tenor of the settlement began to change. The Stuart brothers did a brisk trade in horses and provisions every time a group stampeded after a rumor (which was fairly often), but repeated and increasingly frequent failures in July and August seemed to cast a dark pall over the camp, and violence escalated. Whiskey overconsumption led to fights, at times with gunplay. When another group of horse thieves rode into town in late August, one of them was shot in a local saloon, a second was hung after confessing his crime, and the third was chased out of town when it became clear that he was innocent. According to Stuart, the man in the saloon had gone for his gun when approached by a sheriff from western Idaho during a game of Monte; Stuart reported in his personal journal that the man had died grasping his playing cards so tight that he was buried with them still in his hand. The bucolic and relatively peaceful days the Stuart brothers had experienced during the two previous summers in the valley were over.[14]

In late August and early September 1862, the Stuart brothers heard a series of reports from the Beaverhead and Big Hole Valleys suggesting that placer deposits of greater value than those in Gold Creek had been uncovered far to the south, but the locations and reports conflicted. Granville learned later that a group of gold seekers from Colorado, trying to find a shortcut to the Salmon River diggings across the passes southwest out of the Beaverhead Valley, had happened upon the rich deposits along Grasshopper Creek that would become the first high-volume gold creek mining district in the region. Reports of decent gold prospects along what was then known as Willard's Creek, high above the Beaverhead Valley, had led several prospectors to an almost simultaneous convergence at the location in the midst of Horse Prairie, more than 100 miles south of the Deer Lodge Valley. As dozens of men began working individual claims, the creek bed gradually conceded signs of potentially extensive deposits of gold. Word of the new location quickly made its way to Fort Benton, diverting some members of a group of prospectors and emigrants on their way from St. Paul, Minnesota. The news also spread like wildfire among a few of the smaller gold diggings in the region at Gold Creek, Prickly Pear Creek, and the upper Wisdom River. Before the end of September, most of the men from these locations had abandoned their claims and relocated to the new district.[15]

For dozens of years, the rolling prairie along the winding path of Grasshopper Creek had served as the wintering grounds for a large collection of Bannock, as there was ample graze for horses and enough fuel wood along the banks of the creek for heat. Just as a new kind of political space had been required on the regional level to allow access and assure some level of safety for gold seekers, however, placer creeks also required new and specific kinds of spatial practices to facilitate a relatively safe and regular access to the potential wealth. Like so many western placer creeks since 1849, Grasshopper Creek was organized into a mining district by the men who had begun taking up claims, in this case taking the name from the original inhabitants and naming it the Bannack Mining District. Although formally unrecognized by state, territorial, or federal law, a mining district was a formal agreement among a group of miners in a placer mining location to allocate sections of the creek into regular-size claims, limit individual claim ownership, mandate minimum work required for claim ownership, allocate water rights, and form collectively into an extra-judicial body called a miners' court—composed of the district's miners—which adjudicated all disputes related to the district. Thus, even before Grasshopper Creek had conceded very much gold, the miners had organized its social space to regulate work, allocate claims, and resolve conflicts.

Despite the federal government's expensive work and use of violence in killing Columbia Basin leaders to open the region to exploitation, the state played no roll in the creation of this mining district or any of the hundreds of others that had preceded it in the mining west. Yet the districts were almost universally effective and, because of the way they made property usage rights along the creeks a collective problem, miners adhered to the regulations and to the miners' court decisions as if they had the actual force of formal law. By so doing, disputes were quickly resolved, and most men were liberated to do their work in their claims and to spend time prospecting other locations without fear that they would lose their holdings.[16]

Because Grasshopper Creek had been opened late in the season—it is not unusual for snow to settle in the northern Rocky Mountains as early as September—the arrival of fall inspired an intensification of work and settlement as miners sought to get as much as they could out of their claims before winter set in. As a result, the Bannack District claims yielded a steady washing of gold dust well into the fall. The miners organized commercial space, platting the town of Bannack City in September. During October and November, dozens of additional miners—stragglers from the south or prospectors working their way east from western Idaho—continued to trickle into the growing settlement. Another large overland wagon train arrived from St. Paul led by James K. Fisk, and many of these immigrants came to Bannack City as well. By November 1862, the new settlement held more than 400 people. The Stuart brothers—having entertained Fisk in their cabin and among the last to leave the American Fork settlement in the northern Deer Lodge Valley—disassembled their sluices, packed up their tools, abandoned their Native American wives, and moved to Bannack City, where they opened a butcher shop.[17]

JAMES HENRY MORLEY

On the afternoon of June 20, 1862, after six weeks on a crowded steam-powered paddleboat—a 1,600-mile journey upriver from St. Louis, at times utterly boring but ending tragically with the death of a crew member—James Henry Morley, his partners, and a hundred or so other passengers clambered off the deck of the *Spread Eagle* onto the dry land of Fort Benton, Idaho Territory, at the eastern edge of the northern Rocky Mountains and the navigational head of the Missouri River. They were on their way to prospect for placer gold in western Idaho. During the summer and fall of 1861, dozens of men had trickled out of the west side of the northern Rocky Mountains, down the Missouri River, and into St. Louis with gold dust they had mined along the Clearwater and Salmon Rivers, tributaries of the more substantial Snake River that marked part of Idaho Territory's western border, about 900 miles further west of Fort Benton. Morley and his partners hoped to tap this fortune by staking claims on one of these rivers and mining gold for themselves. James Morley was a civil engineer who had led railroad development through rural southern Missouri in the late 1850s and wanted no part in the escalating violence of the Civil War then under way. The western Idaho gold rush gave him the escape he was seeking. When commercial steamboats offered passage to the northern mountains for the first time in the spring of 1862, he had jumped at the opportunity. Morley had left behind his wife, Virginia, and their six-year-old daughter, Frances, and traded his middle-class urban life in St. Louis for the potential riches of gold in the mountain west.[18]

Morley and his partners found themselves among a sea of prospectors in Fort Benton, as all four of the steamboats the US Congress had commissioned to run the upper Missouri River that year had arrived within days of each other, each packed to the brim with eager emigrants and mining provisions. Amidst the buzz, Morley and his party quickly organized a team with a few dozen other men and secured twenty oxen to pull ten wagons through the mountains across the recently completed Mullan Road. They loaded their provisions and tools, added two additional teams of men who had asked to join them for safety, and slowly rolled out of Fort Benton on a determined plod toward the western Idaho goldfields 900 miles farther west.[19]

They had barely left Fort Benton behind them when they began to encounter significant challenges. Mullan Road, for one thing, was more a wide rutted path than a road, especially compared with the well-used and more carefully engineered roads of Morley's eastern experience. The teams managed a slow and arduous pace, trudging no more than twenty miles a day and usually less. Such exertions cost time and energy, creating enormous appetites among the draft animals. Every evening the men scrambled to ensure that all twenty oxen and the dozen or so horses had adequate graze (or at least something to eat) and potable water; both resources were not always located in the same places or in ample quantities in one place. They also found a scarcity of trees on the arid eastern slope, so finding adequate fuel for cooking presented a frequent challenge. To make matters worse, they were being feasted upon by mosquitoes. "Our greatest discomfort," Morley complained amid all the challenges they faced, "is in the annoyance from the swarms of mosquitoes that infest the whole country in both high and low places." Five days into their journey along the edge of the swollen Sun River, the flood had completely saturated Mullan Road, making it "so muddy as to be impassable with the loads." The teams "were obliged to skirt the foot of the hills," which not only lengthened the journey significantly but also added new obstacles: "Between the hills were marshy spots in which the wheels would sink half up to the hubs, requiring doubling of teams to haul the wagons through," Morley wrote wearily.[20]

Morley did not directly say so in his journal, but the unexpected hardships of overland travel tempered his ambitions to haul all the way to western Idaho and opened him to the possibility of devising a change of plans. He noted in his journal that a gentleman named Mr. Terry, who was traveling east toward Fort Benton, had showed him more than $500 in gold dust (about a pound of the metal) mined from a creek near the Deer Lodge River less than 100 miles further along the road. Morley and his companions also encountered Frank Worden, who was driving six empty wagons to purchase goods at Fort Benton and brought additional news about the Deer Lodge mines. Morley and his team decided to conserve their energy and hedge their bets. They lightened their load, leaving one-third of their provisions with the government Indian agent at the Sun River farm, and pressed onward.[21]

Almost a week later—worn out by the difficult road, needy draft animals, slow travel, and hungry mosquitoes and ready to give up on the western Idaho gold deposits altogether—Morley marveled at the verdant and picturesque Deer Lodge Valley. From Mullan Road, which skirted the valley on the north, the landscape appeared as a wide, grassy expanse of level golden meadow framed by tall snowy peaks on the west and gentle bald hills on the east. The southern reaches of the valley disappeared beyond the horizon, but the area in Morley's immediate view contained promising signs: "saw a heard of some five hundred cattle . . . belonging to Mr. Grant, who has a ranch nearby," Morley wrote. Shortly thereafter Morley pulled his team into American Fork, a tiny settlement with "three or four cabins, occupied by whites and their Indian wives," including Granville Stuart ("an old Californian," Morley noted with some optimism) and his brother James. The men parked their wagons, put the oxen to graze, and climbed aboard horses to ride up Gold Creek and see the mining developments for themselves. "Found here two sluices worked by three men each and others above, being prepared," Morley wrote. "These sluices take out about one and one half ounces each per day of an excellent quality of gold." Rumors of gold and even seeing gold dust along the trail were one thing, but actually seeing gold produced from the muddy remains of a creek bottom was all the evidence they needed. Morley and his men were staying. They set up camp one-quarter mile from the gold-producing sluices and got a good night's sleep.[22]

As would be expected for newcomers to a new mining district, Morley and his men were not certain of what they were doing, and during July they were out-paced by more seasoned miners who arrived after them. After examining the creeks and ravines in the vicinity of American Fork, Morley staked a claim along Rock Creek, which flowed through a drainage to the west of Gold Creek, and began the physically arduous process of developing the property by channeling water and sinking a shaft into the ground until he reached the creek's hardpan bottom. While Morley and his team labored to prepare their claim, a large group of California gold miners arrived at American Fork, took up claims along Gold Creek, and quickly began producing gold dust in their sluices. Morley and his men dug for two weeks more before they found any washable material, but no gold appeared in their rockers after washing. It was now mid-July, and gold seekers kept arriving at American Fork—including dozens of miners who flooded in from Colorado, having heard about the Deer Lodge mines as they were making their way to western Idaho from the south through the massive Snake River Valley.[23]

As more and more men arrived at Gold Creek, Morley extended his own energies and those of his men toward every possible opportunity. He continued to dig in his Rock Creek claim, beginning a new discovery hole, but he also sent two men south into the Beaverhead Valley and two others to examine unclaimed locations on

Gold Creek—in both cases following up on rumors of new gold. In early August, as he continued to turn up nothing in his Rock Creek claim, Morley visited two of the more productive claims along Gold Creek to see if he could learn something of value about finding gold. He staked a claim on the opposite side of the creek from one of them—again to no avail. On August 9 his men returned from their prospecting trip to the Beaverhead Valley with a small amount of gold dust. Their return with gold caused a stampede of Gold Creek miners back to the location described by Morley's men, but no one else could turn up gold there; prospectors soon returned to Gold Creek emptyhanded.[24]

Worried about his chances at Gold Creek, where more worthless claims were turning up than paying ones and his own prospects were coming up empty, Morley decided to try again in the Beaverhead Valley. He had his men pack horses for a ten-day prospecting trip as rumors continued to fly about gold to the south. The men rode overland for three days before arriving at a location called Horse Prairie, whose mines, according to Morley, were "in a bleak, barren looking place, and are not promising." Many other groups of prospectors had already begun to leave the location when they arrived, but Morley and his partners decided to give it a try and within a day struck what appeared to be rich pay dirt. They staked several claims around the location and guarded them for two days, somewhat fearful of Indian attacks as they were the only gold mining group left at the location more than 100 miles from Mullan Road and the Gold Creek settlement. Horse Prairie was a rolling treeless plateau several hundred feet above the southern Beaverhead Valley and a few dozen miles to its west. The creek wound its way across the prairie before dropping through a sharper incline, cutting several well-worn bluffs along the way. As the late August winds swept across the high plateau, Morley and his men decided to head back to Gold Creek for their tools. They soon happened upon a large group of men also on their way back to the Deer Lodge Valley. They joined that group, taking comfort in numbers and keeping their recent discoveries to themselves.[25]

Even with paying claims discovered, or perhaps because of them, Morley and his men found themselves caught up in continued prospecting, as well as some apparently much-needed recreation. On their way back to Gold Creek, the group detoured westward to the upper Wisdom River on the basis of rumors, but they found the creek location already taken up by a group of thirty Colorado miners making about six dollars each a day from their claims, with no working ground left. It was a Sunday, so Morley and a friend ascended to a snowfield a few hundred feet above the Wisdom River claims and entertained themselves for the afternoon by rolling and sliding in the snow. The next night, the whole group camped with a number of Minnesota men they encountered along Mullan Road. The Minnesotans had traveled overland from St. Paul that summer across the Northern Plains in what was called the Holmes Train, and many were trying to decide whether to stay in the region or continue on to Oregon. The issue was decided for many in the group when another large stampede of men, following a rumor of a new gold discovery at Prickly Pear Creek, came upon the camp the next morning. Prickly Pear Creek was located about halfway between Gold Creek and Fort Benton along Mullan Road. Several men from the Minnesota party packed up and joined the stampede. Morley and his men returned to their camp at Gold Creek, packed their wagons, and started back south for the locations they had staked near Horse Prairie. After losing a day when one of their wagons slipped into the river, soaking their belongings, Morley and his men encountered another team of Minnesotans who had split off at Fort Benton from the other Minnesotans they had previously met and who were on their way to a reputedly high-paying placer gold location found near the southern Beaverhead Valley.[26]

So many men and so much movement should have been a warning to Morley and his partners that times were changing quickly in eastern Idaho Territory. Indeed, by the time Morley and his partners arrived back at their Horse Prairie claims on September 8, just two weeks after their initial solitary discovery, it was clear that they should have returned with their tools more quickly or perhaps never have left. More than 300 men had moved into the location, which appeared to be nearly all of the placer miners in the entire vast territory, and they were working placer claims up and down the creek. All of Morley and his partners' claims had been jumped, and they reacted angrily, immediately taking one of them back by force. The other miners along the creek quickly intervened and convinced them to wait until Sunday, when a miners' meeting could settle the issue more rationally. Morley and his men agreed and spent the four days until the miners' meeting prospecting other locations around Horse Prairie, turning up little of interest.[27]

On Sunday, September 14, 1862, the miners of the Bannack Mining District determined by majority vote that Morley and his party were the prior locators on at least four claims along the creek. For the next two weeks Morley and his men channeled water, sunk shafts, and then began lifting pay dirt into a small pile in the center of their most promising claim. Morley's usually descriptive journal entries are clipped during this period. "In mines," he wrote on one of the few occasions when he wrote at all. Then, on Saturday, September 27, almost five months after stepping aboard the *Spread Eagle* steamboat in St. Louis and two-and-a-half months after deciding to place his luck in Gold Creek, Morley and his men washed $172 in gold dust from the dirt in their rockers on their claim along Grasshopper Creek. The next day they spent it all on ten sacks of flour and 125 pounds of salt.[28]

THE "MONTANA" GOLD RUSH

The Bannack Mining District

We would expect that men in a changing frontier region with the potential for gold exploitation would encounter some initial missteps and even defeats, as both Morley and the Stuart brothers had during the summer of 1862. Trying to find placer gold deposits in a region as vast and unmapped as the eastern slope of the northern Rocky Mountains was uncertain business. There were spatial and quantitative uncertainties about the gold itself, whether it was there and, if so, where it was and in what amounts. There were also social and institutional uncertainties related to gold mining as a social practice—in a society of strangers, for example, who can be trusted, and how and by whose authority could the resources in and around productive creeks be allocated? Much of the movement and activity of men in a newly opened region during a gold rush reflected efforts to answer such questions, or at least to create the means to mitigate some of the uncertainty they provoked.

As it would turn out, the placer deposits buried under Gold Creek ended up being far less extensive than the original six speculators and John Mullan had hoped in the spring of 1862. Further, while the organization of a mining district provided something of the kind of quasi-authority needed to deal with criminal activity and rationalized the allocation of claims, it could not temper the frustration and restlessness of unsatisfied gold seekers. Gold production became the victim of its own success, as it drew more men to the Gold Creek location than the ultimate deposit of gold there could support. But with the opening of Grasshopper Creek and the founding of Bannack City (a location that would produce the first significant amount of placer gold from the eastern slope of the northern Rocky Mountains and that, for a time, would be the first capital of the Territory of Montana after May 1864), one would expect an end to the uncertainties as the gold-seeking men settled into their claims and began to undertake the work that had brought them to this isolated place—assured of their security by the mining district and the miners' court and comforted by the continued production of gold. But the founding of Bannack did no such thing.[29]

In fact, it seemed that, rather than evoke a sense of security, the production of gold at Grasshopper Creek produced more anxieties and new uncertainties. Emily Meredith, one of dozens of arrivals to Bannack that fall, remembered a palpable anxiety among the residents of Bannack City when her family arrived. The Meredith family had set out from Minnesota for California in the spring and eventually arrived in Bannack City as part of a large wagon train after months of changing direction and following rumors in the West. "Many anticipated a severe winter because we were so near the top of the range," she recollected. "Others thought the

Indians would not allow us to stay, or that if [enough] white people came in to obviate that difficulty there would be actual starvation. Some thought the mines themselves would be exhausted in two or three months, or at best it would be impossible to work longer than that, until quite late in the spring." The steady production of gold drew dozens of additional newcomers like the Merediths' large party, who continued to arrive and mark out their section of creek in hopes of striking it rich as well. But as the Merediths' experience revealed, this growth was not welcome in the fall of 1862.[30]

More than worry about newcomers' claims rattled the Bannack miners' confidence; they were also concerned about existing claims. Finding gold in one place in a claim did not guarantee it would be found in all places under a claim or in neighboring claims. "The claims above us are paying from $5 to $12 per day, and those below from $4 to $6 per day," Mark Ledbeater wrote to his family back in St. Paul as he started work on his own. "We will have to work hard for two weeks before we can get our sluices up. We may make a good thing out of it, and perhaps not make our board." Other members of Ledbeater's party decided that the risk was too great and that the work was harder and more uncertain than farming, so they left for Oregon before heavy winter snows locked them in.[31]

In fact, it was impossible to know how valuable a gold claim was without continued work. This pressing reality encouraged men to labor steadily whatever the results. James Morley, for example, had continued to perform the backbreaking work of shoveling dirt and rock and stone, building an internal frame for the shaft as it descended, grading ditches and channeling water, building sluices to wash the potential pay dirt, and preventing flooding in the holes being sunk week after week following his initial find in his Grasshopper claims—some weeks washing out gold, some weeks not. If the dirt the miners found was poor or without gold, as Morley's had been at Rock Creek and would be occasionally in his Grasshopper Creek claim, the process would begin again elsewhere on the claim; because gold was deposited randomly, it was not wise to give up too quickly.

Miners worked incessantly, and where gold existed—even in the small quantities deposited in Gold Creek—it produced net results, actually making its way onto the market. The existence of gold dust in the market, a step or two distanced from the labor that had produced it out of the placer creek beds, drew new prospectors and miners with hopes of similar success. The more gold produced by a district, the more gold seekers would be drawn to the location. But as more people were drawn to the location, the faster the location would reach the inevitable limits of its gold deposit. In this way, successful gold production accelerated the ultimate end of the gold rush settlement, creating new uneasiness and uncertainty about the life of the district. Miners never knew when the limits of the present deposits would be reached, only that they would, so the presence of newcomers always represented a potential threat. At least one failed stampede out of Bannack occurred that winter after rumors of a new find over the mountains pushed this uncertainty into action. To make matters even more uncertain in Bannack City, the commercial district had been platted near a regular wintering ground for several large groups of Bannock families, which had arrived in large numbers shortly after the Stuarts moved to town in November.[32]

Despite the uncertainties, the incessant work continued to produce gold from Grasshopper Creek, and the steady production of gold led to a steady flow of immigrants in search of more. Soon, a sizable population of miners and others had piled into Bannack City. In late fall 1862, some of the Minnesotans from the Fisk wagon train decided to see if they could serve the local miners' market by making a run to Salt Lake City, about 400 miles to the south, with empty wagons they hoped to fill with flour and beans and mining supplies. Martha Edgerton, who had traveled to the region as a young girl with her family that fall, had encountered this train on her way into the region, remembering that "more teams were met than formerly" when her group started north toward Bannack out of Fort Hall in the Snake River Valley. "Most of them were on their way to Salt Lake to get winter supplies for Bannack before the season closed for travel," she wrote. But the season did not close the roads in October or November or even December, and the payment of gold dust for supplies in Salt Lake City had caused a stir among the merchants established there. Rumors of a new California somewhere north of Salt Lake City were whispered in every direction along the trails out of town. Through the fall and into the winter, a steady stream of wagons and settlers—in search of the gold dust they saw circulating in Salt Lake City or were shown on the Overland Trail—traversed Corinne Road, bringing flour and beans, mining provisions, and especially the gold seekers who would need those goods.[33]

The uncertainty about growing populations and impending winter conditions that had driven the Minnesotans to run for provisions had set in motion a flow of market exchange and regional population movement that must have come as a relief to the geographically isolated Bannack City residents. They knew that steamboat passage and thus additional supplies from the east would not be possible until the following June and July. But the sudden unexpected increase in wagon traffic along Corinne Road between the mining camp and Salt Lake City led to resource uncertainties for another group of people completely unrelated to, uninterested in, and ultimately hostile to the patterns emerging from the gold production in the northern Rocky Mountains.

A long stretch of the road in southeastern Idaho Territory coursed through the sovereign lands of the Shoshoni people, who had not agreed to any access to or travel

through their lands. Although the wagon trains and miners had no intention of settling and only used this landscape for passage from Salt Lake City to Bannack, the
mode of transportation they used was profoundly consequential for the Shoshoni
who made their lives there. Wagon trains were like locusts on wheels. As had been
the case with Morley's wagon train the previous spring and the dozens like him haul-
ing across open country, every ox team and all of the horses required food and water
every evening—preferably in grazing material—and every person along the way
required meals, often composed of wild game and river fish. As the number of immigrants coursing through the region continued to increase during the fall and into the
winter, the impacts grew in kind and ultimately led to a strong reaction by the sovereign residents as they witnessed the destruction of vital natural resources near the
best camping locations along the road.[34]

In response to these territorial incursions and abuses of their resources, the
Shoshoni acted to defend their lands. In early January 1863, a group of Shoshoni
warriors killed ten miners in an ambush inside their territory, and one week later
they killed two express riders in the same general vicinity. The attacks represented a
generally accepted, if somewhat violent and unexpected, diplomatic response to the
unlawful taking of sovereign resources by an alien people, but respect for the tradi-
tions of international law and practice was never well developed among gold miners.
The attacks challenged the ability of Salt Lake City merchants to exploit the lucrative mining community growing to their north.

The US response to the attacks began at least overtly within the bounds of the
legal system and with gestures toward the rule of law, but the final expression of
revenge ultimately came at the hands of a bloodthirsty US Army colonel. When
word of the killings reached Chief Justice John Kinney of Utah Territory, he ordered
Colonel William Connor—who commanded a ragtag assemblage of infantry and
cavalry fresh from fighting Native Americans in California—to arrest and bring to
trial the men responsible for the murders, preferably the Shoshoni elders in charge
of the group of warriors as well as the culprits themselves. Connor marched north
with about 90 well-armed men and two mountain howitzers, traveling 120 miles to
the banks of the Bear River nestled in the mountains to the south of the Snake River
Valley, where a large extended family group of Shoshoni had set up camp for the winter. There, Connor and his men took up positions surrounding the camp early on
the icy clear morning of January 29. When a few armed Shoshoni men began to take
up a defensive position of their own, Connor ordered his men to open fire. Some
chroniclers of the Indian wars in the US West dispute labeling the battle at Bear
Creek a "massacre," but the loss of 22 of Connor's soldiers compared with at least
250 Shoshoni seems to support its use. However the assault is classified, it brought
an end to Shoshoni attacks along the road connecting Salt Lake City to Bannack by

way of a treaty of peace and the promise of secure passage through Shoshoni territory, which in turn kept the goods and people flowing into Bannack uninterrupted and secured what would become the most important market connection into the region during the early period of the "Montana" rush.[35]

ALDER GULCH AND VIRGINIA CITY

Despite the initial anxieties and uncertainties and the unexpected threat from the Shoshoni, by early 1863 the warm winter, unfettered market connections, and—most important—continued production of gold dust from Grasshopper Creek gave the miners at Bannack a growing level of confidence. Out of such confidence emerged a willingness to take additional risks, a willingness expressed most forcefully by James Stuart, the younger and more adventurous of the Stuart brothers. Even before the rush, James was a notorious gambler, well-known among the cattlemen and Native peoples in the region as a willing player whenever a game of chance was called. James would jump at any opportunity to set aside his household duties to wager horses and goods for nearly any gamble, whether shooting, cards, or one of several games invented by Blackfoot or Flathead men. When miners began to show up in numbers in the spring and summer of 1862, James was more often found in a card game than working the brothers' claim in Gold Creek. James had, in fact, lost a bundle to the card shark and horse thief shot by the Idaho sheriff that summer. With the steady winnings now being pulled from Grasshopper Creek in a region he and his brother had called home for two years, James seemed inflated with the confidence of a gambler who believes he's hit a winning streak. He decided to press his luck further and gamble on a rumor he and Granville had been hearing since first coming to the region in 1857.[36]

According to trappers and mountain men, there were positive signs of a substantial placer gold deposit along the banks of the Big Horn River, a tributary of the Yellowstone River a few hundred miles east and south of Bannack City. Granville and James had first prospected Benetsee (now Gold) Creek on the basis of similar rumors from the same kinds of men, and that investment had ultimately come to fruition, albeit not exactly as they might have intended. The main obstacle to exploiting the Yellowstone location prior to this time was the troubling fact that the Big Horn River was well beyond the bounds of the treaties with Native Americans that had opened the northern Rocky Mountains; in fact, it was deep into what everyone clearly understood was Crow territory. The Crow were no more likely to approve of such a violation of their sovereignty than the Shoshoni had been. But James Stuart believed a large contingent of armed gold seekers could change the equation. By the beginning of April 1863, he had convinced almost two dozen men to rendezvous in

Routes Out of the Gold Rush Region

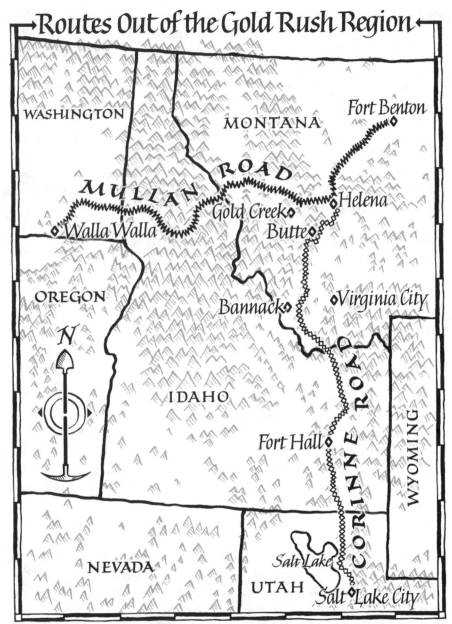

Until the end of the US Civil War, only two major routes coursed through the gold rush region of the future Montana. Corinne Road was the busier of the two. The Bear River flowed to the east of Corrine Road in southeastern Idaho. Corrine Road represented the most important market route into the region during the early gold rush period. Original art by Ruth Pettis.

the Ruby Valley, east of the Beaverhead Valley, to make an armed expedition to the Big Horn River location and see what they could turn up.[37]

The expedition did not begin exactly as planned, but, all things considered, James was not disappointed. The first fifteen men arrived at the agreed-upon rendezvous location between April 9 and April 11. After securing horses and signing an agreement to follow all of James Stuart's orders, the group pressed on to Stinking Water Creek further down the Ruby Valley, where Stuart had arranged to meet six additional men. "We can find no trace of them," Stuart wrote after the group arrived at the location. "They have failed by some causes unknown to us." The group camped nearby for two more days, hunting elk and exploring the drainage while they waited, but the tardy men failed to arrive. To make matters worse, two men from the expedition party found gold in a sandbar along the river; Stuart made them promise to keep the discovery quiet "for fear of breaking up the expedition." The next morning the group struck camp and headed toward the mountain divide that separated the Madison River from the Yellowstone River. This journey took twice as long as Stuart had anticipated, a circumstance he blamed on the map he and his men had followed: "Lewis and Clarke have played us out; if we had left the notes and map of their route at home and followed the Indian trail, we would have saved four days' travel." Nevertheless, they found the Yellowstone River and began following it eastward toward the Big Horn.

After another week of travel sullied by rain and snow and bad-tasting water, the men were overtaken by a group of Crow scouts, who tried to steal their horses and supplies and stayed around the camp perimeter for twenty-four hours. As the expedition continued eastward, the Crow scouts remained close, looking for an advantage. The Crow tried once again to wrestle the horses and supplies from the Stuart party, but this time Stuart and his men drew their rifles and threatened the group leader's life, which seemed to shake the Crow from their intentions. Finally, on May 4, three weeks after leaving Stinking Water Creek, the thirteen-man Yellowstone Expedition arrived at the Big Horn River.[38]

Having witnessed the haphazard development of both Gold Creek and Bannack City, James Stuart sought to arrange for an orderly development of what he hoped would be his new gold mining community along the Big Horn River. He and his men immediately found encouraging signs of gold near the mouth of the river. On the second day, Stuart sent four men further upriver to continue prospecting while he and five others platted the new development of Big Horn City, a 640-acre town site surrounded by thirteen 160-acre ranches, each assigned to one of the expedition members. With the new town platted, Stuart and his men began making their way south and further upriver toward the Big Horn Mountains far in the distance, to continue their prospecting work and identify additional mining districts for which they would hold the discovery claims. As they continued southeast along the river, they noticed increasing signs of nighttime visitors and occasionally caught sight of a few mounted Indians far in the distance, leading them to post two nightly guards to prevent what they feared would be more attempts to steal their horses.[39]

Unfortunately, James Stuart's hopes for an orderly, easy development of a new gold mining community along the Big Horn River came to a tragic and violent end less than a week after the city was platted. On the cloudy, dark evening of May 12, the horses appeared spooked when Stuart and George Smith began their watch. Both men decided to lie prone on the ground near their horses to see what was making them nervous, suspecting it might be a wolf. Stuart reported: "Just then I heard Smith whisper that there was something around his part of the horses, and a few seconds later the Crow fired a terrific volley into the camp. I was lying between two of my horses, and both were killed and very nearly fell on me. Four horses were killed, and five more wounded, while in the tents two men were mortally, two badly, and three more slightly wounded."[40]

As quickly as the shooting began it stopped, but fearful that the shooters remained and would open fire again, Stuart shouted to his men to pull down their tents and take up defensive positions away from them. The men remained silent through the rest of the night except for the moaning of the wounded. At daybreak, they gathered up the necessary supplies and a few surviving horses and decided to head due south toward the Overland Trail. One of the seriously wounded men died in camp that morning, and the other asked to be left behind with his loaded pistol. It would take the men almost six weeks to make the roundtrip south, west, and then back north to Bannack City; several of the men decided not to return to Bannack. By the time James Stuart and his remaining men limped back into Bannack City on June 22, bedraggled and bearded such that "scarcely anyone knew us at first," they learned, with what must have been growing dismay, that their miscalculations had been even greater than they imagined. As they would forever regret, the men realized that they should not have left their camping location in the Ruby Valley and Stuart should not have suppressed the information about the promising discovery of gold there prior to their departure for the Big Horn River.[41]

It would be left to the six men who failed to catch the Yellowstone Expedition to uncover the next great placer gold location in the region, although not without a series of missteps of their own. The men had originally arrived at the Stinking Water Creek location almost two weeks early, but their horses ran off in the night and it took the men several days to track them down. They decided, after having a difficult time finding the Beaverhead River, to try to circle ahead of Stuart and the expedition party and wait for them to pass. Unaware that Stuart was following the notes and maps of the Lewis and Clark Expedition, the men had camped above the valley

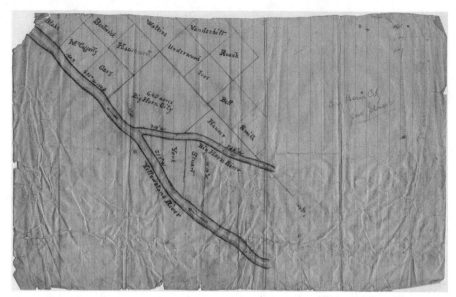

Map of "Big Horn City," May 6, 1863. The plat is oriented toward the southeast and was drawn up five days prior to the attack on the Yellowstone Expedition's camp. As this map reveals, experienced gold seekers knew the real money in gold rushes was made through real estate. As the gold settlement grew, the value of these initial holdings would have skyrocketed. James Stuart Diaries, courtesy, Montana Historical Society SC 1877, Helena.

where they expected the expedition to pass and waited almost two weeks. On April 15, Henry Edgar and another man rode into the valley and found horse tracks along a part of the river they had not been watching; they were certain the tracks were from the Stuart party. For the next two weeks the group attempted to follow the trail, finding and losing and then finding it again as they traveled.

Just as they believed they had almost caught up with Stuart's group on May 1, they noticed many more tracks than those left by the Stuart party: "Looks as if there were thousands of them," Edgar wrote, "and the worst is that they have been here since Stuart passed." Confirming Edgar's fears, the men were overpowered and taken to a Crow village in the midst of a celebration of a successful buffalo hunt. They were held as captives for three days (increasingly fearing they were about to be killed) before being given worn-out horses and set on the trail back to the Ruby Valley. Five of the six men took the offer and retreated as quickly as possible toward Bannack (the sixth man decided to stay with the Crow, some of whom were his relatives); they were told that if they tried to continue after the Stuart party, they would be killed. To ensure their compliance, twenty scouts followed them all the way back to the pass out of the valley. On May 14, just two days after Stuart and his men had been shot up at the Big Horn River, Edgar and the others finally left the Yellowstone Valley and reached the Madison River: "We are safe," Edgar wrote in his journal.[42]

Grateful for their lives, the men began to make their way back to the original rendezvous location near the juncture of Stinking Water Creek and the Ruby River, prospecting creeks along the way. The men found many locations where they believed gold might be found, but they never panned quite enough gold to cause them to keep the locations. By the time they arrived at Stinking Water Creek on May 24, their horses were exhausted and nearly lame. The men decided to rest a day or two before continuing on home to Bannack City. Rest, in placer gold country, meant parking the horses and prospecting ground, which the five men did in earnest.

Little did they know that their lives were about to become very interesting. Henry Edgar and William Fairweather, who had stayed in the lower gulch to watch the horses while the other three explored the creek above, discovered a rim-rock—an exposed piece of bedrock along which one could easily reach potential pay dirt—and sampled for gold. Much to their surprise and delight, they recovered almost $12 in gold after a few pans. The next day all five men worked the location with pans and washed what became increasingly rich pay dirt, collectively washing about $175 in gold from pans alone during the day. The following day the men staked out twelve locations running 100 feet down the fall of the creek, named the location Alder Gulch, wrote up paperwork about water rights, and made plans to return to Bannack City. "We agreed to say nothing of the discovery when we get back to Bannack and [to] come back and prospect the gulch thoroughly to get the best," Henry Edgar wrote in his journal.[43]

The men believed they could sneak into town, get what they needed, and return to the location to prospect further to gain an advantage regarding the quality of different locations; however, they had no idea how much Bannack City had changed during their two-month absence. When the men had left for the Yellowstone Expedition in early April, there might have been as many as 700 men in Bannack City. As the spring settled across the West, the mining settlement experienced a renewed surge of newcomers responding to the steady flow of gold that had made its way out of the region all winter long. James Morley watched this population growth firsthand along Grasshopper Creek. "A large arrival of Colorado miners today," he wrote in his journal on April 24. Less than three weeks later, on May 12, the same day the Yellowstone Expedition was attacked in their tents, "A train of twenty-two wagons came in, mostly from Colorado and miners."[44] As had happened at Gold Creek the summer before, gold seekers arriving in the mining district soon began to press the limits of the gold deposits. "Poor show here for the great crowd," Morley noted. One week later, "Street now alive with men, so great has been the influx of people." By early June, Bannack City seemed suddenly headed for a greater overpopulation crisis than had occurred in Gold Creek in 1862. In the eight weeks since the Yellowstone Expedition had left Bannack, the mining settlement had grown nearly six times over, to an estimated 4,000 people.[45]

When Edgar and company returned to Bannack City to buy mining provisions, horses, and new clothing with sacks of gold dust, they were not returning to the same community they had left in April. Despite their promises to each other to keep the find quiet, they could barely suppress their excitement. "Friends on every side," Edgar wrote. "Bob Dempsey grabbed our horses and cared for them. Frank Ruff got us to his cabin. Salt lake [sic] eggs, ham, potato. Such a supper!" While the men revealed their information only to a few select friends, their behavior and that of those close to them raised great interest in the mining city. "Got what we wanted and were all ready for the return, but it is impossible to move without a crowd," Edgar wrote. Morley also noted the stir about the town. "Quite an excitement to go to new gold mines, said to have been discovered not over two hundred miles from here and many are prepared to follow the discoverers back," he wrote on June 2. The creek location was in fact only sixty-five miles to the east, but the discoverers did not want to risk having their claims jumped by giving away the location, and they probably hoped to discourage some by exaggerating the distance. The following day Morley noted, "A large crowd left in a.m. for the new 'diggings.'"[46]

Granville Stuart also encountered the stampede as he returned from a supply run to the Deer Lodge Valley. "They were strung out for a quarter of a mile," Stuart wrote with amazement at the spectacle, "some were on foot carrying a blanket and a few pounds of food on their backs, others were leading pack animals. The packs had been hurriedly placed and some had come loose and the frightened animals running about with blankets flying and pots and pans rattling, had frightened others and the hillside was strewn with camp outfits and grub." Edgar was less descriptive and probably more annoyed by the outcome: "A crowd awaits us—crowds follow after us—they camp right around us, so we can't get away."[47]

While the men had hoped they could sneak back out of town, it became clear that no such thing was possible and that to avoid losing access to the grounds they had already claimed, they would have to develop a more organized strategy. On the first night out of Bannack, the men and several hundred followers stopped and set up camp at the foot of Beaverhead Rock, where Edgar and his companions called a meeting. The men admitted for the first time that they had discovered what they believed to be a significant deposit of placer gold, and they insisted that everyone who had joined them recognize their prior claims in perpetuity. If not, the men would not lead them to the location. "Some talk and it was put to a vote," Edgar wrote. "The vote was in our favor; only one vote against it." The crowd insisted on being told the location after the vote, "but as some were on foot and others on horse-back with the advantage," Edgar and the others decided it was best to withhold this information until they had arrived close to the bottom of the creek.[48]

Two days later, the morning after camping just below Stinking Water Creek and after surreptitiously sending two men ahead to ensure that their stakes were not pulled by any unscrupulous gold seekers in their midst, Edgar announced to the crowd that they had arrived. "Such a stampede!" Edgar wrote. "I never saw anything like it before." Gold seekers piled over one another as they rushed up the embank-ment to stake claims on their own potentially rich slice of the creek. Henry Edgar stepped clear of the frenzy, helped at least one man find a location where he and the men had panned a valuable washing of gold a week earlier, and watched the Alder-lined drainage fill up with men, horses, and boundary stakes. By late afternoon, not a single stretch of the creek remained available; every last inch had been claimed. Three days later the miners organized the Fairweather Mining District, through which they agreed on how claims and locations would be allocated and held and how disputes would be adjudicated.[49]

The Fairweather Mining District at Alder Gulch quickly emptied Grasshopper Creek and Bannack City, creating an instant settlement and all the opportunities and challenges that represented. "People flocked to the new camp from every direc-tion; all the other settlements in the country were deserted," wrote Granville Stuart, who moved to Alder Gulch with James at the beginning of July to open a blacksmith shop. "There were no houses to live in and not much in the way of material to con-struct houses," Granville remembered. "Every sort of shelter was resorted to, some constructed brush wakiups [sic]; some made dug-outs, some utilized a convenient sheltering rock, and by placing brush and blankets around it constructed a living place; others spread their blankets under a pine tree and had no shelter other than that furnished by the green boughs overhead. The nearest sawmill was seventy-five miles away on the creek above Bannack."[50]

While the stampede in early June attracted the surplus men, those whose claims seemed less promising or who had no claim, and all of the newest arrivals in the region, others like Granville Stuart and James Morley—who already had a signifi-cant investment in Grasshopper Creek—were less quick to gamble on the new loca-tion. Granville Stuart not only waited with some concern about his brother, but he had also just returned to Bannack City from a supply run for his butcher shop and grocery and decided to continue to serve the miners who remained with their claims in the Bannack District. However, when James returned empty-handed at the end of June, the brothers quickly sold off their last provisions and moved to what had become Virginia City, the major settlement and commercial center of the Fairweather Mining District. Morley had also waited, continuing to work his claims and wash his pay dirt even as the first stampede nearly emptied Bannack of its excess miners during the first week of June. When word came in mid-June that the Alder Gulch mines were paying well, news that pulled many successful miners from their Grasshopper Creek claims, Morley continued to work his ground. Not until the end of the month, when news arrived that the new diggings were continuing to prove fairly extensive, did Morley finally give in and head over to take a look.[51]

A year of work in the placer mines had tempered Morley's anxiety about new claims and sobered his reactions to the unusually violent social and work environment that existed in placer gold communities. Morley was a changed man. He and his partners arrived at Alder Gulch just as two men had been hanged by the relatively new mining court. When Morley had first landed at Fort Benton a year earlier, he had witnessed the drowning of a deck hand who was trying to help pull the steamboat up the final stretch of rapids below Fort Benton. This episode was described in great detail in his journal, with a deep sense of sorrow at the loss of human life. In July 1863, Morley mentioned the hanging of men with a detached, matter-of-fact tone, and he paid much more attention to his own fate than to the hanging corpses he rode past. "No chance," he concluded about his odds of finding an unclaimed mining location in Alder Gulch. He and his party sullenly watched Fourth of July fireworks in Virginia City and then spent the rest of the month prospecting locations around the outer perimeter of the Alder Gulch claims, in surrounding creeks, and in other unclaimed ground in the vicinity without any luck; it seemed that all the paying locations had been found. Morley grew so desperate that, for the first time since coming to the region, he worked on a Sunday. "Thinking of the wants of loved ones at home," he justified, "it seems no sin, in this savage country, to exert oneself on their behalf, on the Sabbath." Because of the large number of gold seekers who had flooded into Bannack City in early 1863, the wave of seekers anxious to secure a claim along Alder Gulch was both larger in size and closer in proximity to the new diggings than the year before, resulting in an even more rapid shortage of available paying claims. The scarcity seemed to have put everyone on edge right away, as the July 15 miners' meeting ended in a fistfight. By the end of the month, Morley and his partners decided to do what most eager miners were doing in the gulch: they purchased an existing claim from a speculator and hoped that it produced as well as its owners had declared, spending the gold they had mined in Grasshopper Creek to gain access to potential gold in Alder Gulch.[52]

The social conditions were not the only thing that had changed. By moving from Grasshopper Creek to Alder Gulch, Morley had placed himself in a new and different set of geological conditions. "These claims are very hard to open," Morley complained in his journal after four solid weeks of work. "The 'bed-rock' line lies some twenty feet under the surface, the material being all loose gravel, through which the water of the creek flows freely and as the creek has a grade of only some three to four feet per hundred, it is necessary to cut a drain ditch five or six hundred feet long to get to the rock." "It is discouraging to fight so much water," Morley concluded, "but I am used to fighting discouragements." Despite his attempts at fortitude and courage, the difficulties kept coming. Morley had to purchase a mechanical pump to keep the steadily seeping groundwater from filling his discovery shaft, but, to his chagrin, the pump kept filling with sand and other debris that washed down from claims upstream and required frequent repairs. When they finally did reach bedrock, the waters ceased flowing in the creek altogether "owing to sluicing operations three fourths of a mile above." By mid-October, Morley and his partners' patience was wearing thin, but a miners' meeting on October 11 relieved the pressure to work, allowing all claims in the district to go unrepresented until the following April. Morley and most of his men returned to Bannack City to perform the necessary work to maintain possession of their Grasshopper claims, while his brother and the others remained in Alder Gulch and built them a cabin near the new claim.[53]

Despite the challenges and discouragement faced by men like Morley and James Stuart, the net product of almost 10,000 gold seekers digging shafts and washing pay dirt was a steady and growing stream of gold coming out of the northern Rocky Mountain region. This development was of significant interest to the Union government and to Secretary of War Henry Stanton, who sent a second military escort, led again by James Fisk, during the late summer of 1863. The region was rapidly being populated by southern men, miners from Georgia and Missouri who had flocked to Colorado in 1859 and 1860 and were now abandoning that territory en masse. This strong presence of southerners potentially sympathetic to the Confederacy led the members of Fisk's first wagon train to develop their own settlement, called Yankee Flats, at the edges of and across the creek from Bannack City. The following year (1863), as the mining town associated with Alder Gulch was being incorporated, most miners in that location wanted to name it "Varina City" after the wife of Confederate president Jefferson Davis. It was the unilateral decision of one of the miners' judges in recording the vote that gave it the name "Virginia City," a more palatable compromise to the southerners. With an estimated $10 million in gold dust having been produced out of Grasshopper Creek, Alder Gulch, and several other smaller placer locations by the end of 1863, these developments could not have been a minor concern to the Union government, whose ability to maintain a slight advantage over southern forces during the year could easily have been tipped the other direction by the influx of specie into Confederate coffers.[54]

Fisk's original plan had been to drop the emigrants at Fort Benton and then to proceed along Mullan Road to examine its conditions all the way to Walla Walla, but as his wagon train approached Fort Benton he learned about the gold developments along Alder Gulch and decided it would be worth his time to explore the two mining settlements that had sprung into existence since early 1862. He stopped first in the Prickly Pear Creek gold diggings, which he reported had "not yielded much as yet." Even after more than a year of development, the men were still hard at work trying to dam and ditch enough water to wash their pay dirt. Further along Mullan Road toward the Deer Lodge Valley, his men stopped to wash dirt along the Blackfoot River in a spot that had been prospected and worked until June 1863, when its miners had left for Alder Gulch. Fisk was even more impressed than Morley had been the season before when he arrived in the verdant Deer Lodge Valley, where he found the Deer Lodge River to be "a fine stream of pure water" and noted John Grant's 4,000 head of fattened cattle and 3,000 horses. Fisk spoke glowingly about the valley, calling it "an admirable tract for grazing and farming":

> Wheat and oats grow luxuriantly at Dempsy's farm, and vegetables of all kinds are raised. The grass is sweet and excellent, and there is fine timber on the mountain sides . . . Grant's cattle range the valley the whole winter; many of the animals are so fat that their appearance is similar to that of Berkshire shoats fed for the fair. Some of my party of 1862 left work-cattle here in the fall that were thin and worn out with the journey across the plains; in April they were very fat, and were sold for beef cattle.[55]

Fisk visited Gold Creek but found that it, too, had been abandoned, as all of the miners seemed to have flocked to Alder Gulch by the fall of 1863. He then headed south through the Deer Lodge Valley and crossed over to the Big Hole Valley, where he noted two additional gold digging locations at Wisdom River and Rattlesnake Creek that had also been abandoned for Alder Gulch. Finally, he made his way south and east to Grasshopper Creek and the Bannack City mining settlement.[56]

Fisk's report on his visit to Bannack City and the Grasshopper Creek diggings is instructive both to understand the evolving social and political conditions in the new gold mining region and to gain insight into how quickly a creek bed could be transformed by the work of placer miners. Fisk and his men announced their arrival under the US flag by firing their rifles into the air as they approached the settlement. Nathaniel Langford, a military man himself and one of Fisk's assistants during the 1862 Fisk Train from St. Paul, rushed out to meet Fisk and his men with a large group of additional miners. The large crowd gave three raucous cheers to the American flag and then escorted the men into Yankee Flats, the large settlement of Union sympathizers headed by Langford. "We seem to have met a band of brothers," Fisk remarked using a military description derived from Shakespeare, "so kindly and hospitably were we received." Fisk noted little else about the sectional tensions present in the landscape geography, but his careful rendering of the developments and production of gold at Bannack City—which was written up and forwarded to the

US Department of War in Washington, DC, shortly after he returned to St. Paul—shows how important this information was to the military.[57]

Along Grasshopper Creek, Fisk described finding placer mining activities that had effectively turned the creek basin and surrounding landscape into a busy rural factory. He described four miles of claims along the creek, each "fifty feet wide, stretching from base to base of the mountains." "These are the 'Poor Man's' mines," Fisk explained, "requiring little capital to work them." Here, several hundred miners washed an average of ten dollars a day in gold per man from claims, often using water that had been ditched to the location from the tops of mountains ten or fifteen miles away from the creek. While "some rich claims producing large amounts" of gold dust were still being worked along the creek, in many places miners had excavated all of the dry ground away from the main body of the creek and begun diverting the creek itself to mine underneath its existing bed.[58]

Of more interest to Fisk than the rapid decline of the placer diggings were the slowly emerging quartz outcroppings that had been claimed along the hillsides and canyons above the creek bed. Only the ores from what had been named the Dakota Lode had been worked in a makeshift stamp mill constructed out of wagon parts and using wooden blocks as stamps, but they showed promising results. So far, the quartz had produced a wide range of yields from week to week. Sometimes a week of crushing and washing produced $2,000 in gold; sometimes the same volume of ore produced a mere $300 in gold, although the mine owner believed his ores were becoming richer as his mine went deeper. It was difficult to say for sure, though, as the mining work had all been done with a pick and shovel alone, and the miners had not yet cleared through the brittle, weathered surface ore to the more solid and usually more consistent quartz below. Fisk noted that around fifty other quartz claims had been located along the hillside above Grasshopper Creek, all of which had been preliminarily opened with shallow discovery shafts and most of which had sampled the ore within, with many samplings as rich as the ore mined from the Dakota: "I am confident that when the quartz lodes of Bannock [sic] are properly worked, the yield of gold will be *permanently* greater than that of the Virginia City mines." Fisk also speculated that many more quartz lodes like the ones he visited in Bannack would soon be found in other parts of the same mountain range. He was encouraged by the potential for nearby agriculture to support such endeavors. "I learned that the valleys of the Jefferson, Madison, and Gallatin forks of the Missouri were good agricultural districts," he wrote. "Grain and vegetables can be raised there. Grass and water are good, and there is an abundance of pine timber on the mountains." Before he left Bannack, Fisk learned from the US marshal that Mullan Road had been completely closed to passage by deep snows, so travel to Walla Walla would be impossible.[59]

He traveled next to the Fairweather Mining District. Despite the fact that Virginia City and the Alder Gulch diggings were less than four months old when Fisk visited, the settlement already contained several thousand people and had outgrown the original plat for Virginia City, spilling into what had been named Nevada City. The settlement consisted of "a long street of stores and cabins, with side streets branching off at right angles. The general appearance is the same as at Bannack City," Fisk wrote. The more rapid appearance of a larger settlement resulted from the richer placer deposits excavated from the creek bed. They averaged twenty dollars per miner per day, according to Fisk's estimates, about double the yield of the Grasshopper Creek deposits; many Alder Gulch claims produced significantly more gold than that, "as high as $50, $75, and $100 to the man daily," Fisk reported. As Morley had experienced, these were claims of a higher order of difficulty than those along Grasshopper Creek, requiring an investment of between $200 and $1,500 to dig, pump, and frame a drift under twenty to twenty-five feet of creek detritus to access the potential pay dirt buried below—"but, when once opened, the claims have invariably paid well." Fisk estimated that 8,000 miners worked a fifteen-mile stretch of Alder Gulch leading from Stinking Water Creek up to Virginia and Nevada Cities, which among them had produced about half a million dollars in gold every week since midsummer. The Fairweather Mining District's claims were twice as wide as those at Grasshopper Creek, at 100 feet along the creek and from ridge to ridge. Similar widths were being applied to three new nearby creeks—Harris's, Brown's, and Bevis's Gulches—which were just being opened during Fisk's visit. Based on what he saw, Fisk anticipated significant growth in placer mining output in the year ahead.[60]

In an interesting way, the experience of the gold rush outcomes on the eastern slope of the northern Rocky Mountains depended on one's angle of approach and the perspective from which it was viewed. Morley's work and that of thousands like him in the region did not always produce results, was at times dangerously frustrating, and seemed to become exponentially more expensive after the move to Alder Gulch. By October 1863, in fact, Morley's earnings from his Bannack claims were being poured into his Alder Gulch claim without any return on the investment. James Stuart had invested two-and-a-half months and fifteen lives—losing three of them along the way—in his dream of creating a new mining settlement at the Big Horn River, but instead of beginning a ranch on the new plat of land he had laid out, he went to work as a blacksmith in Virginia City that fall.[61]

The individual stories were uneven, contingent, uncertain, but when one swooped in, as Fisk had done, and took a regional snapshot, the results appeared much more certain and impressive. Fisk estimated that more than 12,000 miners were working claims in the Bannack and Virginia City mines by October 1863, which meant that in the eighteen months since the 6 prospectors first arrived at Benetsee Creek, the mining population had grown to equal or exceed the indigenous population living in the region before that time. According to statistics Nathaniel Langford provided to Fisk, these miners were unearthing between $600,000 and $700,000 in gold dust every week, and Fisk estimated by his own observations "that nearly $15,000,000 of gold [was] . . . waiting [for] safe transportation to the eastern States." Thus the cumulative story, having ironed out the uneven individual experiences, presented an impressive set of positive statistics. Perhaps out of sheer arrogance or perhaps to serve a Union claim on the goldfields of the northern Rocky Mountains, Fisk took credit for these impressive results: "It is a source of the highest gratification to me, that such should have been the amazing results of the discoveries made in the fall of 1862 by the party of emigrants in my charge, and under the protection of the government expedition." Before the end of October 1863, Fisk sold the remainder of his provisions and horses and boarded an express wagon in Bannack for the long, bumpy ride back to Salt Lake City.[62]

As Fisk made his way south, he passed multiple wagon trains of miners who had left Colorado's goldfields for the new northern mines. At the Snake River, Fisk reported seeing "one hundred and fifty wagons from Denver" on their way to the new mining region, and before Fisk and his team reached Salt Lake City they had "met about four hundred teams from Colorado, and learned from the emigrants that that Territory would be almost depopulated by the immense emigration." Fisk made no remarks about the Coloradans' political leanings, but it is easy to suspect that this news needed no qualification and probably left Secretary Henry Stanton concerned enough about the miners' sympathies to continue sending Fisk to accompany northern emigrants over the next several years. As they passed through the mountains to the south of the Snake River Valley, the wagon team made a detour to the location along the Bear River where William Connor's troops had killed the Shoshonis the previous January. "Many of the skeletons of the Indians yet remained on the ground, their bones scattered by the wolves," Fisk wrote with cold-hearted approval. "Since their punishment the Snakes and Bannocks have made a treaty with the general, and not one emigrant has been molested."[63]

Fisk's detour and comments on the massacre were intentional and directed to a clear military policy he believed should be expressed in the region. James Stuart had shared with Fisk the experiences of the Yellowstone Expedition at the Big Horn River and requested a new military post at that location to protect the miners from Indian attack—in a sense, an extension of a mining frontier eastward from the present boundaries. But Fisk asserted the need to follow a more violent and forceful approach than the negotiations that had rendered the northern Rocky Mountains available after 1859: "No one thing, in my mind, has ever proved so fatal and disastrous to our march of western settlement as the sending of commissioners and blankets, in advance of emigration, for the purchase of territory." Instead, Fisk wanted a swift and aggressive campaign, like the one executed by Connor, to take the remaining sovereign indigenous territory in the US West "with impunity." "The Indian cannot appreciate kindness," Fisk declared. In a revealing justification, Fisk suggested that the entire vast region of the Yellowstone River Valley, then violently defended by the Crow, contained at least $200 million more in gold deposits. Connor's swift massacre at Bear River had helped put $10–$20 million worth of gold into circulation; a similar region-wide effort was sure to liberate the rest of these treasures. Indeed, given Fisk's accounting of Connor's growing military forces in the region around Salt Lake City and his encouragement that his men prospect when they were not on duty, it appeared that he believed Connor should take charge of such a campaign. Thus the cumulative results of the eighteen months of accidents and uncertainty on the ground led Fisk to imagine an unproblematic victory against the Crow and an unproblematic exploitation of regional gold resources.[64]

Like many boosters of western development, Fisk's vision suffered from too much abstraction and too little attention to context, place, and contingency. In his calculation, the only obstacles to safe and regular gold production and the subsequent circulation of gold specie in the economy were hostile groups of indigenous people preventing access to reserves. He not only overlooked the radical uncertainty that was at the material foundation of placer mining work itself, as experienced by Morley and Stuart, among others, but he also neglected to recognize the uncertain social dynamics that would soon erupt from successful gold production in the northern Rocky Mountain landscape.

The tensions and mistrust that existed among a growing mass of gold seekers and that were exacerbated by the opposing ideologies of the Civil War, from which these northern mining camps did not escape, were made exponentially more complicated as successful miners and, more often, successful merchants moved to bring the gold they had secured out of the dense settlements and the region. Montana's unique topography and the geography of its first two major gold creek deposits in relation both to each other and to other, more settled regions of the country forced miners and merchants to traverse extensive unsettled and unpopulated distances composed of wide, open valleys linked by narrow canyons and isolated mountain passes. Throughout the fall of 1863, as miners and merchants prepared to move more than $10 million in gold around and out of the territory, these in-between places became spaces of uncertainty. Because of the nature of gold dust, it was impossible to differentiate one person's gold from another's; if a more criminally minded person succeeded in gaining control of someone else's gold, it would quickly become impossible to identify its source. Gold's universal qualities combined with its cultural

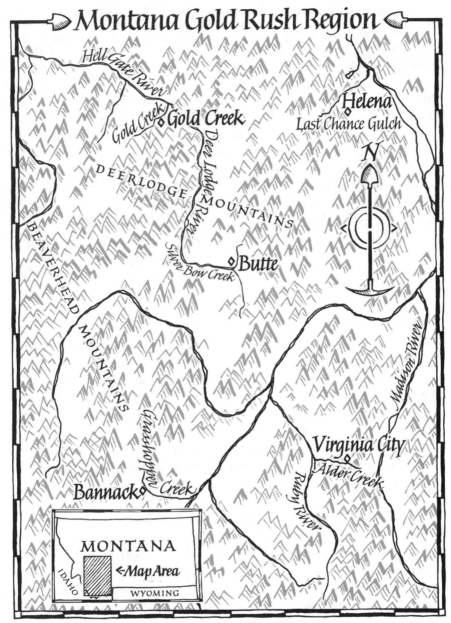

Montana Gold Rush Region

Hell Gate River

Gold Creek ◇ **Gold Creek**

Helena
Last Chance Gulch

DEERLODGE

Deer Lodge River

MOUNTAINS

N

BEAVERHEAD

Silver Bow Creek

◇ **Butte**

MOUNTAINS

Madison River

Virginia City
◇
Alder Creek

Grasshopper Creek

Bannack ◇

Ruby River

MONTANA

◀ Map Area

IDAHO

WYOMING

This map shows the relative distances among the three major placer gold deposits in the region that would become Montana Territory in 1864, as well as Gold Creek, which launched the region's gold rush, and Butte, where a small amount of placer mining would later give way to significant lode mining. Original artwork by Ruth Pettis.

characteristics to make it a generally dangerous commodity to try to move in large volume across open country.

As the cumulative successes began to pile up in the hands of merchants and among the handful of unusually successful claimants during the 1863 mining season, people in the mining settlement expressed increasing fear about overland travel and worried about the material and perhaps physical losses that could occur. When Fisk and his small group of advisers had first made their way toward Prickly Pear Creek along Mullan Road, for example, he reported to have frightened a miner on his way to Fort Benton. The miner had become concerned that Fisk and his men were thieves intending to rob him of his gold—Fisk and his men had changed from their military uniforms into buckskin suits before leaving Fort Benton—so the miner had tossed his gold far into the brush before approaching the group. When the mistake was discovered, Fisk and his men spent time helping the man search for his satchel, which was eventually recovered and found to contain $3,000 in gold dust.[65]

Such fear and uncertainty must have followed many of the miners who sought to bring their gold back East. Morley noted a robbery in October 1863, as did Granville Stuart. Merchants were equally intimidated as they sought to move even larger volumes of gold. Most significant, Nathanial Langford described being followed in early November by a group of men as he and Samuel Hauser hauled $10,000 in gold dust belonging to him and the Stuart brothers for deposit in Salt Lake City. Langford and Hauser arrived safely and returned without incident. In a book published many years later, Langford wrote of multiple stage robberies and murders during these months, with impossibly detailed recountings of dialogue and exchanges during events at which he was not present. Morley's daily journal notes that he met with Langford once or twice during the fall of 1863 about his trip to Salt Lake City, but he makes no mention of any robberies other than the one in October and mentions no murders on the trails. Granville Stuart's edited diary mentions a murder that he believed stemmed from a robbery in late November, after Langford and Hauser's return, to which he added: "There has [sic] been several robberies on the road between here and Bannack seventy miles away and few are people living between the two camps. There is certainly an organized band of highwaymen about here and something will have to be done soon to protect life and property."[66]

A continued historical controversy rages over the legitimacy of the events that next unfolded in and around the mining camps in December 1863 and January 1864. It seems fairly certain, however, that the combination of Crow resistance, sectional conflict in a distant region, a huge volume of gold dust in need of safe transport, and a struggle for the exercise of authority over the rapidly growing new settlement and the vast spaces in between all conspired to create one of the most notorious examples of what has been called "frontier justice" in the history of the US West.

Finding among the many strangers several members of the Order of the Masons (including Samuel Hauser and James Stuart), Nathaniel Langford, himself a Mason, determined to identify and eliminate what he and other Masons believed was an organized gang of highway robbers targeting successful miners and merchants as they attempted to leave the region with their gold. Their suspicions led them first to George Ives, one of the survivors of the ill-fated Yellowstone Expedition in 1863.

The group of vigilantes, which included Langford and Hauser (who had also survived that expedition), held a secret trial in which Ives, they claimed, proceeded to name almost two dozen men as co-conspirators—including the current sheriff of Virginia City and southern sympathizer Henry Plummer, who Ives is said to have fingered as the leader of the group. Ives was then executed by hanging. After the turn of the year, the vigilantes determined the locations of the twenty other men Ives had named as co-conspirators and proceeded to systematically and summarily execute them. They hung two men along Stinking Water Creek on January 4; five men, including Plummer, at Bannack City on January 10 and 11; four men in Virginia City on January 14; three men at their separate winter camps on January 16, 19, and 24; four men at Hellgate near present-day Missoula on January 25; one man at Fort Owen on January 26; and the last reputed member of the gang in the Gallatin Valley on February 3. Based on the fact that the preponderance of evidence proving these men's guilt was assembled only later in a narrative form by the same man who had led the vigilantes—Nathaniel Langford—and has usually been attributed to one or more of the executed men as they were killed, it is impossible to determine satisfactorily whether these executions, and the half-dozen or more that followed during the rest of the year, were in any way justified by events.

It is certain, however, that like the mining districts themselves, they were fundamentally illegal, without formal sanction of law, and completely absent of any of the necessary institutional procedures designed to make the rule of law dependable and secure in the United States. Yet none of the vigilantes was ever tried (Langford, in fact, was appointed the first superintendent of nearby Yellowstone National Park eight years later).[67]

In the mining camps, these actions appear to have been met with approval, providing a sense of security in an otherwise uncertain landscape. "Since the summary proceedings of the Vigilance Committee order and quiet prevail throughout the country and one feels as safe, if not safer, than in the States. The miners are a well-disposed class and are all intent on getting fortunes," Morley explained. If murder was necessary to serve their intentions, then so be it. In the same way James Fisk approved of the violent actions General Connor took to rid the regional transportation routes of threats to the safe movement of goods, specie, and miners, local miners at Bannack and Virginia City appeared to view the vigilantes' violent work as

Samuel T. Hauser, late 1870s. Courtesy, Montana Historical Society Photography Archives, catalog #940-409, Helena.

a necessary exercise of gold community protection.[68]

In what may not have been coincidental timing, shortly after the vigilante actions and after two years of rapidly expanding gold production and population growth, the men in the two major mining communities on the eastern slope of the northern Rocky Mountains voted to apply to become a recognized territory of their own, including all of the mountains east of the Bitterroot Range as well as a huge swath of the northern Great Plains as far east as the juncture of the Yellowstone and Missouri Rivers. A petition was sent to the US Congress to consider accepting a new territory named Montana, to be carved out of eastern Idaho and parts of western Dakota and Nebraska Territories. On May 26, 1864, at the encouragement of Sidney Edgerton—who had ventured to Washington, DC, to lobby President Abraham Lincoln and the Congress—the legislation was passed and Montana Territory officially came into existence as part of the Union.[69]

Montana Territory

The $20 million in gold that had circulated out of the region by early 1864 not only engendered violent efforts (and calls for renewed violence) to keep gold and goods moving and to liberate new potential gold regions, it also generated a similar exponential influx of gold seekers who believed the region must hold opportunity for them as well. But the same overpopulated conditions experienced in Gold Creek and Grasshopper Creek during the past two summers became part of the landscape of Alder Gulch and Virginia City in the spring of 1864. "Emigration continues to flow in and as there have been no new mines found, labor is abundant and many are disappointed. 'Pilgrims' find gold cannot be found in every stream by merely the trouble of picking it up, an idea many have until they come and learn for themselves," Morley wrote, perhaps reflecting on his own long labor over the past two years in an effort to achieve moderate success. While there were no new mining claims around

Territorial governor Samuel T. Hauser, circa 1886. Hauser had arrived in the new gold region with the first Fisk Train in 1862. His driving ambition and shrewd alliances helped him rise from gold seeker to territorial governor. As was the case with many of Montana's early leaders, Hauser's rise to power took shape after he particpated in the 1864 vigilante campaign and invested early in Helena real estate. Photograph by R. H. Beckwith; courtesy, Montana Historical Society Photography Archives, catalog #940-409, Helena.

either Bannack or Virginia City for the new miners to take up, those with some capital to invest began enlisting new strategies to earn something from the territory's gold production. In both of the established locations, the need for water and other forms of mining construction (e.g., opening underground shafts, building and assembling stamp mills, constructing flumes to transport water) generated a steady demand for additional skilled and wage workers. Morley, for example, reported that he hired laborers to work in his Bannack claim that year. Those without the skill or the inclination to work on construction usually headed out to prospect elsewhere in the region.[70]

There was never another gold deposit in the region bigger than the one found at Alder Gulch, and there would never be another substantial find as easy to access as either of the first two discoveries, although by the end of 1864 miners had begun to develop deep placer deposits at Last Chance Gulch (and had founded the city of Helena) and had finally begun to wash some gold at the Prickly Pear Valley deposits Fisk had visited in 1863. A few dozen miners began to work another new group of gold-enriched bars along Silver Bow Creek in the southern reaches of the Deer Lodge Valley.

The 1864 mining season experienced intensive exploitation rather than expansive new discoveries. Because the work of getting the gold was even more difficult in many of these new locations, they required more labor and thus represented greater work opportunities for the large numbers of eager gold seekers. In Last Chance Gulch, pay dirt was more than thirty feet under the surface, often with less of a slope (diminishing the creek water's energy capacity). Equally often, enriched gravel was found in dry creek beds, requiring water flumed from elsewhere. For these reasons, the new placer mines required more work and the construction of more extensive infrastructure to become successful gold production locations. The need for sluices was good news for men like Morley, whose surveying and civil engineering services were increasingly needed and who found more men willing to pay for them. But it was difficult for merchants like the Stuarts to keep up with the demographic and market shifts; they remained in Virginia City too long to take advantage of Helena and saw their profits slip and their business fail. By 1865 the new placer mines outside Helena were poised to contribute significant gold dust to the regional and national economy and to replace the dropoff in gold production experienced by the exhaustion of the richest deposits at Grasshopper Creek and Alder Gulch.[71]

The continued increase in gold production in Montana Territory during 1865 continually drew more men to the region. Helena quickly became more populous than Virginia City. Because of its location near Mullan Road and its proximity to shortcuts south into the placer mining districts, Helena became the supply center for a vast region of smaller gold districts in Jefferson, Powell, and Lewis and Clark Counties, in addition to the expansive material demands of the Last Chance Gulch

Virginia City, Montana, Alder Gulch in foreground, 1866. Just three years after its founding, Virginia City was already experiencing a significant decline in population, as large-scale hydraulic mining was the only potentially remunerative form of mining that remained in the gulch. Photographer unidentified; courtesy, Montana Historical Society Photography Archives, catalog #956-061, Helena.

deposits and several other gold bars being washed west of Helena. By summer 1865, Sidney Edgerton, a future governor of Montana Territory, estimated that as many as 10,000 people were "overflowing" the new placer gold locations around the new city, where "thousands of men are there taking out the gold."[72] That same summer, after selling his claims in Bannack and Alder Gulch and looking at a few prospects around Helena, James Morley climbed aboard a steamboat and returned to his middle-class life in St. Louis, having netted a little more than a thousand dollars in gold after three years of hard work. The Civil War had ended that April, and Morley's departure marked a significant demographic turn as men who had been avoiding the war began returning home at the same time men who had been fighting the war began flooding the West.[73]

By 1865 the older placer mining creeks—called colloquially "Poor Man's Diggin's"— in Gold Creek, Grasshopper Creek, and parts of Alder Gulch were running through ugly heaps of stone and sand. Whatever valuable pay dirt remained in these places would need some form of large-scale hydraulic mining or diligent teams of Chinese miners to be extracted. Companies of men invested in the construction of long flumes to transport water dozens of miles over mountain ranges and across ridge lines into "dry" gulches around Helena, and several developers began the practice of hydraulic mining above Grasshopper Creek, by which they washed entire banks of sand through ground sluices—all with decent returns. The new energy- and labor-intensive mining practices kept the territory's gold production at a steady rate through 1865, the peak year of gold production in Montana, but the days of the independent miner and small prospecting teams were coming to a close. The investment in infrastructure (capital) gave the owners of the large-scale technologies (ditches and quartz mills) growing dominance over the character of the production process, and work was increasingly organized along industrial patterns of specialized jobs and divisions of labor.[74]

The shift from individual and partnership claims to the consolidation of larger and larger sections of districts, combining multiple claims into a huge single claim worked by vast mass-production techniques in the industrialized rural landscape of western Montana, attracted a new class of workers to the regional mining effort after 1865. The advent of mass production was yet another hedge against the uncertainty mining presented, as it used an economy of scale to offset the lower-value deposits being worked. "It is not a country for miners, in my judgement [*sic*], at least not like Cal. in early times," a disappointed William Taylor wrote to his sister Eliza after arriving in Montana Territory in July 1866. By "miners" he clearly meant skilled laborers like Morley, not the workers who took his place. "There are scattered over the country several places where gold has been found in large quantities—but these are from 25 to 150 miles apart, and then only three or four make money. Helena is in a little narrow ravine something like Hangtown—crowded with people, all asking for work, and hard to get." The available gold mining jobs at this point would have consisted almost entirely of work on other men's claims at wages significantly lower than the averages of 1864 and 1865 and certainly far lower in value than the gold actually being extracted by that labor.[75]

By all estimates, Montana gold production peaked in 1865; while the territory's production continued to be second only to California throughout the rest of the decade, it fell off steadily after 1866. Ironically, however, the movement of goods and people into the region continued. Disconnected from the actual experiences on the ground and perhaps still hoping that a new rich set of placer gold deposits would be found somewhere in the region, commercial steamboats continued to find willing passengers to pay for the ride upriver—perhaps largely as a result of the pent-up ambitions of young men who had served in the US Civil War and were now seeking opportunities in the West. The continued influx also suggested the growing demand for heavy equipment and mining supplies, stamp mills, explosives, and drills. An average of thirty-two steamboats a year docked at Fort Benton between 1866 and 1870, with a high of thirty-seven boats making the long trip upriver in 1867. In 1868, according to one estimate, thirty-two steamboats collectively hauled almost 5,000 tons of freight and as many as 6,000 passengers to or near to Fort Benton. These numbers are especially impressive when one considers the relatively short window of about two months in which steamboat passage was possible to Fort Benton in each of these years. But no more rich placer gold creeks existed in the region, and the gold rush days were over. By the early 1870s, gold production in Montana Territory fell into a precipitous decline.[76]

THE LEGACY OF THE GOLD RUSH

Montana's gold creeks would yield more than $100 million in gold dust between 1862 and 1870, significantly less than the estimates made by Fisk in 1863 but a substantial production nonetheless. Knowledge of this figure and the fact that the gold it measures was once resting in creek beds around the state gives Montana's gold rush a certain inevitable feel. If gold was there, men would find it.

But gold seekers like Morley, Stuart, and thousands of others experienced the gold rush as anything but inevitable. Instead, driven by anxiety and uncertainty, gold seekers had scoured their way across the northern mountains. Men tested every likely creek and sought promising locations, although more often than not they turned up nothing. Gold seekers also followed rumors—some of which had long swirled in a region and others that had been generated by recent newcomers. But it was the not knowing that pressed prospectors and gold seekers into the mountains and valleys of the US West and across the Territory of Montana—not knowing if there was more gold to be had, if other hidden riches were buried somewhere. Proving a negative is a difficult task, even more difficult than working a paying claim. So once signs of a new deposit of gold appeared, everyone who heard the news or saw the evidence raced to the location like cattle in a stampede, desperate to gain access to a new, potentially valuable claim. Extra-legal social structures ordered this desire into something other than a free-for-all, and gold seekers submitted to the rules layered atop the creek by mining district regulations. But even the certainty of the miners' courts could not determine if a claim was indeed valuable. Placer miners had to spend significant time digging and washing to know for sure whether their claim contained gold or to determine that it did not. Morley was not behaving foolishly in 1862 when he tried for weeks to make his Rock Creek claim pay or when he stayed in Grasshopper even after many others had abandoned their claims for Alder Gulch; instead, he was being responsible.[77]

At the district level, uncertainties about the actual extent of the creek deposits made the arrival of newcomers the source of social anxiety. As long as new claims continued to prove up, growth in population was accompanied by growth in the market, stimulating a profitable commercial center and robust trade. Successful gold miners attracted ambitious traders and drew sufficient goods to the new marketplace in gold settlements. But a steady influx of gold seekers, as happened year after year in Montana and elsewhere, quickly led to an exhaustion of paying claims and created a growing body of frustrated gold seekers. The inability to stake paying ground in a district awash in gold dust led many idle miners to seek other means to gain access to the gold. Some invested in the lucrative business of commercial trade, others played endless games of chance with the men who did have gold, and still others turned to theft and robbery. Because there was a delay between the time of locating claims and the realization of the limits of the gold deposits, the risks of overpopulation were always present in a nascent form in gold rush communities. This created a paradoxical tension in the settlements, where men worked to access unknown quantities of gold within a landscape whose limits would only be discovered by being exceeded; it was a social order bound to create deep frustrations. It is no wonder that placer gold settlements were so notoriously chaotic and persistently violent.

The paradox, however, derived not from the uncertainty alone but also from the intermittent certainty that made gold a part of American society. In certain times and places it was found, sometimes in large amounts. Just as the gambler is drawn to the game not by the devastating losses that inevitably come but because sometimes—in random moments of existential benevolence—he actually wins, gold seekers tolerated the uncertainty that troubled every gold rush settlement because in times and places again and again across the US West, gold had been deposited in large amounts. The game always proceeds on the possibility of winning, and the gold rush period in the US West proceeded according to the same rules. For a finite period in various places across the mountain west, men cleaned their rockers and sluices at the end of a hard week of digging and found gold dust there in the riffles. They could pay for another week of work or more.

The possibility of success combined with the uncertainties about the value of one's own claim, as well as the existence of unknown other locations, generally drove men to work existing claims as quickly as possible and to be willing to try out new locations, so stampedes and frenzies were a regular part of gold mining settlements. The behavior of gold miners, which is largely attributable to the layers of uncertainty on which placer and gold mining was built, nevertheless produced a cumulative result—seen best by outsiders like James Fisk—that was itself a paradoxical and ironic marker of success. In the first place, as more gold was produced in the places where it had been deposited, market activities and increased migration intensified the impacts on transportation routes into and out of the main camps, causing localized success to have a broad geographic reach and often requiring the ever-widening exercise of social control to secure the safe and relatively risk-free movement of men, goods, and gold. In Montana, as elsewhere in the West, this social control was often expressed as violence and frequently included extra-judicial assaults and murders of Native people and accused criminals. Such violence was not incidental to the overall gold-seeking process in the US West; because of its unique spatial and social character, it was in fact part of the gold rush package from the foothills of the Sierra Nevada to the Black Hills of South Dakota.

A less obvious irony also accompanied the successful working of a valuable placer gold deposit: namely, that the frenzied production of gold was at the same time the frenzied advance toward the certain exhaustion of any particular deposit. Just as miners could not know the value of their claim until they worked it and mining districts could not know the extent of their deposit until after it was reached, the actual knowledge of gold came at the expense of the gold deposit itself: the more a miner knew he had by having extracted it, the less he actually had in the ground. The faster gold was excavated, the more rapidly the gold deposit was eliminated and a district was rendered valueless. It was a material irony no creek district could avoid. For Montana, despite steady immigration and growth throughout the 1860s, peak production had been reached in 1865, and all of the placer diggings uncovered had become effectively worked over by the early 1870s. Success and hard work in the diggings had brought about an end to the placer mining locations. The social response, however, was continued population growth for another half decade, under more and more tenuous circumstances.

Because of the speed with which the region boomed and busted around placer gold production, historians of Montana's mining industry have not dwelt long on these episodes. They are seen as "colorful" and "curious" but understood as hardly significant to the real mining that would come later. Isaac Marcossen's 1957 company history of the Anaconda Copper Company called the gold rush years "a picturesque chapter in the saga of Anaconda."[78] Michael Malone, the dean of Montana mining history, committed a mere six pages to the gold rush era in *The Battle for Butte* and focused most of his attention on Silver Bow Creek, one of the era's less substantial placer mining locations and important only because of its proximity to the Butte silver and copper ore deposits. "Everyone knew that the placers must soon fade," Malone concluded, "and fade they did."[79] More recently, Timothy LeCain's environmental and technological history of western mining doesn't even mention the gold rush period, and Fredric Quivik's environmental history of copper mining in Montana omits it as well.[80] Thomas Rickard, America's first mining historian, did credit the gold rush with pioneering mining in Montana, but in the end he merely chronicles the region's successful gold stampedes during the period 1862–64 as an unproblematic series of "discoveries." Like Malone, he gives unwarranted emphasis to the Silver Bow gold diggings before quickly turning his focus to lode mining on Butte Hill.[81] Given the character of gold rush society, as well as the radical shift in scale that came with capital investments after 1880, it is easy to see why the gold rush era might be seen as a quaint and insignificant introduction to the real mining action that came later. It is much more accurate, however, to assert that it was *because of* the character of the gold rush society that Montana became a valuable and important mining region.

This argument is easier to make if we conceptualize the gold rush and its impact as a system that has emerged out of related but nonlinear activities rather than as discrete and linear expressions of simple cause and effect. From this systems perspective, parts of the story that we've already seen can be understood as elements of input, output, and flow; and the movement of information and materials can be seen as components of this system's feedback mechanisms. The existence of Mullan Road and the availability of steamboat travel and military escorts across the Great Plains did not determine that a gold rush would occur in Montana, but they did channel a vast array of individuals (who were, incidentally, hoping to reach the western slope of the Rocky Mountains in present-day western Idaho) who had multiple, complex ideals and ambitions toward a concentrated arrival in the vicinity of Benetsee Creek in the summer of 1862.

The visible production of gold in that location served as one of the more powerful positive feedbacks to gold seekers. Because of its hidden and uncertain existence, the visible production of gold led the larger group of gold seekers to invest their time and energy beyond the actual existence of gold deposits, always drawing more men than the location could accommodate and pushing miners to labor in their claims long after rockers and sluices ceased yielding valuable quantities of gold dust. Under these social conditions, rumors of gold became almost as powerful as the visible production of gold itself, so the gold mining society was marked by intermittent stampedes that led to nothing. In the terminology of systems theory, gold and rumors served as powerful forms of positive feedback information (loops that tell a system to keep doing what it is doing) while the absence of gold took much longer to act as a necessary form of negative feedback (information loops that stabilize a system). The result, of course, was the rapid discovery and exploitation of Montana's several placer gold deposits and the seemingly ephemeral rise and fall of placer gold mining in the region over the course of five years.

Viewed through the lens of systems, however, the failed stampedes and rushes are of equal importance; in fact, it is critical to explore the impacts of the system

Map of the mountain section of the Fort Walla Walla and Fort Benton military wagon road from Coeur d'Alene Lake to the Dearborn River, Washington Territory, constructed under the direction of the US Department of War by Capt. John Mullan, US Army; surveyed and drawn by Theodore Kolecki, CE, 1859–63. Size of original map 53 × 127 cm. In the section between the two forks of the road in the lower right it says "Mountains partly covered with pine timber. Unexplored," just to the north of the label "Gold Mines" near Gold Creek and Deer Lodge Valley in the lowest right beneath the south fork of Mullan Road. Courtesy, Montana Historical Society Archives, call no. A-63, Helena.

as a whole rather than simply narrate the material results of its waxing and waning gold production, A system cannot generally be understood according to its feedback loops alone but rather should be seen as the overall result of its cumulative responses to those loops. From this perspective, the gold mining society did something much more permanent and valuable to a nascent mineral industry than simply produce $100 million in gold dust: it produced knowledge. Through its steady, chaotic, and complex reactions to the strong positive feedback and the weak negative feedback, the gold mining society mapped the broader mineral potential of the northern Rocky Mountains; it produced the blueprints for the development of a mineral region.[82]

Nothing more clearly reveals this production of knowledge than a comparison of two actual maps constructed at either end of the Montana gold rush. In 1863, using knowledge acquired between 1859 and 1862, John Mullan created a map of the region through which the Mullan Road had just been built. In this map, Mullan Road makes a singular detailed course through a vast, blank territory. Gold Creek alone is marked with the promising suggestion "Gold Mines" and a nearby label stating "mountains partly covered with pine timber unexplored" (see map of the mountain section of Mullan Road). In 1865, W. W. de Lacy assembled and drew the first official map of Montana Territory with knowledge acquired between 1862 and

Map of the Territory of Montana, with portions of the adjoining territories, showing the gulch or placer diggings actually worked and districts where quartz (gold and silver) lodes had been discovered up to January 1865. Drawn by W. W. (Walter Washington) de Lacy for the First Legislative Assembly, 1865; courtesy, Montana Historical Society Archives.

1865. In de Lacy's map, drawn just two years later than Mullan's map, every major gold deposit is located. In addition, the map contains a wealth of information about quartz lodes and other promising leads, which peppered the details of the mountains with potentially lucrative gold and silver quartz locations. There are no blank spots on de Lacy's map; even James Stuart's yet-to-be-occupied Big Horn City is included (see map of the Territory of Montana).

Through the uncoordinated behavior of the gold rush system, an area the size of southern New England became a known mineral region. An informal geography of minerals emerged from this behavior, producing knowledge that was formalized by de Lacy in 1865. De Lacy's map attests to the effectiveness of this system: gold seekers had not only turned up all of the productive creeks that would ever be found in the region, they had also *eliminated* many more locations as well as noted and identified dozens of potential ore lodes. The importance of such knowledge production was not lost on federal mining commissioner Rossiter Raymond, who wrote optimistically in his *American Mining Journal* (soon renamed the *Engineering and Mining Journal*) in 1868 that "within the short space of nineteen years we have opened up to settlement a larger area of territory than has ever before been brought within the range of civilization in the same length of time. It was a search for precious metals that first carried the adventurer away from the culture and comforts of an eastern home, across the plains, over the Sierras, until at last he struck upon the gold shores of the Pacific. Nor has the end yet come."[83]

In spite of Raymond's hopes, the gold rush system's incredible efficiency in opening the West to settlement could not prevent the region's creeks from running out of gold. In three short decades, placer gold rushes had created the Territories of California, Nevada, Colorado, Idaho, and Montana—each of which would become a valuable natural resource base for the victorious North and the future United States. The incessant push back against uncertainty that animated gold seekers produced invaluable geographic knowledge, suggested new layers of mineral potential in some places, and effectively ruled out that potential in other places. The gold rushes helped to map many of the valuable quartz leads in a region and launched a new culture of natural resource exploitation that would do everything in its power to manage the conditions of inorganic stochasticism and social uncertainty that riddled mining practices. But the gold rush system could not prevent the western placer gold creeks from running out of paying gold. It could not prevent its own self-created and inevitable demise. Indeed, by 1870, if mining was going to continue to support the economy of Montana Territory, it would have to be constructed on a new material basis.

NOTES

1. Malcolm J. Rohrbough, *Days of Gold: The California Gold Rush and the American Nation* (Berkeley: University of California Press, 1997), 126. On the gold rush opening of the West, see Paul, *Mining Frontiers*.

2. Morse, *Nature of Gold*; Smith, *Mining America*.

3. David I. Groves et al., "Secular Changes in Global Tectonic Processes and Their Influence on the Temporal Distribution of Gold-Bearing Mineral Deposits," *Economic Geology and the Bulletin of the Society of Economic Geologists* 100, no. 2 (March 2005): 203–24; Richard J. Goldfarb et al., "World Distribution, Productivity, Character, and Genesis of Gold Deposits in Metamorphic Terranes," in Jeffrey W. Hedenquist et al., eds., *Economic Geology One Hundredth Anniversary Volume: 1905–2005: Society of Economic Geologists* (Littleton, CO: Society of Economic Geologists, 2005), 407–50.

4. R. H. T. Garnett and N. C. Bassett, "Placer Deposits," in Hedenquist et al., eds., *Economic Geology One Hundredth Anniversary Volume*, 813–44. Rickard, *A History*, details the

placer gold locations whose exploitation brought the first miners into each of the major mining regions of the American West after 1848. Rodman Paul, *Mining Frontiers*, argues that these kinds of deposits represented the pull factor on an easterly moving frontier leading out of California's Sierra Nevada foothills in the 1850s. Also see Groves et al., "Secular Changes," 212.

5. David A. John, Albert H. Hofstra, and Ted G. Theodore, "Part 1: Regional Studies and Epithermal Deposits," *Economic Geology: A Special Issue Devoted to Gold in Northern Nevada* 98, no. 2 (April 2003): 225–34. For a helpful chart and discussion of orogenies, see pp. 227–29.

6. Morse, *Nature of Gold*, 22.

7. Albert L. Hurtado, *Indian Survival on the California Frontier* (New Haven, CT: Yale University Press, 1988); Elliott West, *The Contested Plains: Indians, Goldseekers, and the Rush to Colorado* (Lawrence: University Press of Kansas, 1998); Kent Curtis, "Producing a Gold Rush: National Ambitions and the Northern Rocky Mountains, 1853–1863," *Western Historical Quarterly* 40, no. 3 (2009): 296.

8. Curtis, "Producing a Gold Rush," 278–81.

9. Paul C. Phillips, ed., *Forty Years on the Frontier, as Seen in the Journals and Reminiscences of Granville Stuart, Gold-Miner, Trader, Merchant, Rancher and Politician* (Lincoln: University of Nebraska Press, 1977), 133–40.

10. Curtis, "Producing a Gold Rush," 290–92. See also Captain John Mullan, *Report on the Construction of a Military Road from Walla-Walla to Fort Benton* (Washington, DC: Government Printing Office, 1863).

11. Mullan, *Report*, 49–50. The estimated number of non-Indian residents was calculated by listing all of the individuals and parties mentioned in Granville and James Stuart's diaries from 1861 and early 1862. See Philips, *Forty Years*, 157–204.

12. Philips, *Forty Years*, 191, 205–10; Mullan, *Report*, 33–34.

13. Philips, *Forty Years*, 215–20

14. Ibid., 215–29.

15. Granville Stuart, "Montana as It Is," in Paul C. Phillips, general ed., *Sources of Northwest History no. 16* (Missoula: Montana State University, 1931), 5–6; Philips, *Forty Years*, 224–26, 232.

16. Charles Howard Shinn, *Mining Camps: A Study in American Frontier Government* (New York: Alfred A. Knopf, 1948 [1884]); William E. Colby, "The Origin and Development of the Extralateral Right in the United States," *California Law Review* 4, no. 6 (1916): 439–40; Robert W. Swenson, "Legal Aspects of Mineral Resources Exploitation," in Paul Gates, ed., *History of Public Land Law Development*, Public Land Law Review Commission (Washington, DC: US Government Printing Office, 1968), 709.

17. Edwin M. Stanton, Secretary of War, to James L. Fisk, May 29, 1862, Correspondence and Miscellaneous Papers, undated, 1861–62, James Liberty Fisk and Family Papers, P592

Manuscript Collection, Minnesota Historical Society, Minneapolis; Helen McCann White, ed., *Ho! For the Gold Fields: Northern Overland Wagon Trains of the 1860s* (St. Paul: Minnesota Historical Society, 1966), 26–35; Philips, *Forty Years*, 231–32.

18. James Henry Morley, "Diary of James Henry Morley in Montana 1862–1865," 1–17, Small Collection 533, Montana Historical Society Archives, Helena [hereafter MHS]; Virginia F. Morley, *James Henry Morley, 1824–1889: A Memorial* (Cambridge, MA: Riverside, 1891), 19–21, 26–28.

19. Morley "Diary," 17–18; Tom Stout, ed., *Montana: Its Story and Biography* (Chicago: American Historical Society, 1921), 177–80; James McClellan Hamilton, *From Wilderness to Statehood: A History of Montana, 1805–1900* (Portland, OR: Binfords and Mort, 1957), 147.

20. Morley, "Diary," 19–21.

21. Ibid., 21–25.

22. Ibid., 28.

23. Ibid., 29–36.

24. Ibid., 36–40.

25. Ibid., 42.

26. Ibid., 42–48; White, *Ho*, 23–24.

27. Morley, "Diary," 48–49.

28. Ibid., 50–51.

29. Michael P. Malone and Richard B. Roeder, *Montana: A History of Two Centuries* (Seattle: University of Washington Press, 1976), 76.

30. Emily R. Meredith, "Bannack and Gallatin City in 1862–1863," in Paul C. Phillips, ed., *Historical Reprints: Sources of Northwest History no. 24* (Missoula: Montana State University, 1933), 4.

31. White, *Ho*, 45 (Ledbeater quotes); Morley, "Diary," 51–65.

32. Morley, "Diary," 52; Phillips, *Forty Years*, 225.

33. "Autobiography, pp. 101–200," 111, Folder 1–12, Box 1, General Correspondence, 1863–1936, Writings (Autobiography, Lost with Lewis and Clark), Manuscript Collection 78, Martha Edgerton Plassman Papers, MHS.

34. West, *Contested Plains*, details the resource conflicts that arose during the Colorado gold rush in 1859–60 when thousands of wagon trains crossed the Great Plains in hopes of securing a gold claim in the Pikes Peak region. See esp. 145–70.

35. Brigham D. Madsen, *The Shoshoni Frontier and the Bear River Massacre* (Salt Lake City: University of Utah Press, 1985); Gregory Michno, *Encyclopedia of Indian Wars: Western Battles and Skirmishes, 1850–1890* (Missoula, MT: Mountain Press, 2003), 108–11.

36. Phillips, *Forty Years*, 218, 221, 234.

37. James Morley noted that Stuart attempted to have all of the men incorporate as a single entity, but disagreements about the form of that corporation led them to decide to proceed individually on the expedition. Morley, "Diary," 91–92.

38. James Stuart, with notes by Samuel T. Hauser and Granville Stuart, "The Yellowstone Expedition of 1863," in *Contributions to the Historical Society of Montana, with Its Transactions, Acts of Incorporation, Constitution, Ordinances, Officers and Members,* vol. I (Helena, MT: Rocky Mountain Publishing, 1876), 149–53, 160 (quote about Lewis and Clark), 162–80 (remaining quotes).

39. Ibid., 180–90.

40. Ibid., 191–92.

41. Ibid., 192–233.

42. Henry Edgar, "The Journal of Henry Edgar," in *Contributions to the Historical Society of Montana, with Its Transactions, Officers and Members,* vol. 3 (Helena, MT: State Publishing Company, 1900), 132, 134, 135 (quotation).

43. Ibid., 137–39.

44. Morley, "Diary," 95, 100.

45. Ibid., 101; James L. Fisk, *Expedition of Captain Fisk to the Rocky Mountains: Letter from the Secretary of War in Answer to a Resolution of the House of February 26, Transmitting Report of Captain Fisk of His Late Expedition to the Rocky Mountains and Idaho,* 36th Congress, 1st Session (Washington, DC: Government Printing Office, 1864), 27.

46. Edgar, "Journal," 140; Morley, "Diary," 106.

47. "Henry Edgar Journal, Discovery of Alder Gulch," 47, Henry Edgar Papers, 1/1, SC276, MHS.

48. Edgar, "Journal," 141.

49. Ibid., 141–42.

50. Phillips, *Forty Years,* 263.

51. Ibid., 248–57; Morley, "Diary," 107–11.

52. Morley, "Diary," 111–15.

53. Ibid., 117–19; Phillips, *Forty Years,* 263–64.

54. White, *Ho,* 18; Phillips, *Forty Years,* 263; Stout, *Montana,* 199–200; Fisk, *Expedition,* 30.

55. Fisk, *Expedition,* 30.

56. Ibid., 25–27.

57. Ibid., 27.

58. Ibid., 27–28.

59. Ibid., 28 (original emphasis).

60. Ibid., 29.

61. Morley, "Diary," 117–23; Phillips, *Forty Years,* 257.

62. Fisk, *Expedition,* 29–30.

63. Ibid., 31.

64. Ibid., 34.

65. Ibid., 23.

66. Morley, "Diary," 125–26; Phillips, *Forty Years,* 262.

67. Thomas J. Dimsdale, *The Vigilantes of Montana: A correct History of the Chase, Capture, Trial, and Execution of Henry Plummer's Notorious Road Agent Band* (Butte, MT: W. F. Bartlett, 1915 [1865]), 78–161; Nathanial Pitt Langford, *Address Delivered before the Grand Lodge of Montana at Its Third Annual Communication, in the City of Virginia, October 8, A.D. 1867* (Helena, MT: "Herald" Book Establishment, 1868); Nathaniel Pitt Langford, *Vigilante Days and Ways: The Pioneers of the Rockies, the Makers and Making of Montana, Idaho, Oregon, Washington, and Wyoming,* vol. 1 (New York: D. D. Merrill, 1893), 305–400.

68. Morley, "Diary," 157–58.

69. Malone, *Montana,* 72–73.

70. Morley, "Diary," 166.

71. Ibid., 169–76; Stout, *Montana,* 210–13; Merrill Burlingame and Kenneth Ross Toole, *A History of Montana* (New York: Lewis Historical Publishing, 1957), 125–26.

72. Letter from Sidney Edgerton to Sister Martha, June 6, 1865, Outgoing Correspondence to Martha Wright Carter (sister-in-law), 1864–65, Folder 1–5, Box 1, Manuscript Collection 26, Sidney Edgerton Papers, MHS.

73. Morley, "Diary," 176.

74. A. K. Eaton, assisted by William S. Keyes, William de Lacy, and others, Section 3, "Notes on Montana," in Rossiter W. Raymond, ed., *Mineral Resources of the States and Territories West of the Rocky Mountains* (Washington, DC: Government Printing Office, 1869), 145; Rossiter W. Raymond, *Silver and Gold: The Mining and Metallurgical Industries of the United States, with Reference Chiefly to the Precious Metals* (New York: J. B. Ford, 1873), 261.

75. "William Taylor to Eliza Taylor, July 10th, 1866, Helena, Montana," Folder 1866, May–July, William Taylor to Eliza Taylor, Mitchell Family Papers, 1858–87, Henry E. Huntington Manuscript Collection, San Marino, CA [hereafter HHMC].

76. "Steamboat Arrivals at Fort Benton, Montana, and Vicinity," in *Contributions to the Historical Society of Montana, with Its Transactions, Act of Incorporation, Constitution, Ordinances, Officers and Members,* vol. I (Helena, MT: Rocky Mountain Publishing, 1876), 317–25; Raymond, *Mineral Resources,* 138–39.

77. On gold production, see J. Ross Browne, *Report on the Mineral Resources of the States and Territories West of the Rocky Mountains* (Washington, DC: Government Printing Office, 1868), 511. The statistics published by Browne for 1862–67 are disputed by gold production statistics given by William Andrews Clark in his address to the Centennial Exhibition in Philadelphia in 1876. See Hon. William Andrews Clark, "Centennial Address on the Origin, Growth and Resources of Montana," delivered at the Centennial Exposition, Philadelphia, October 11, 1876, in *Contributions to the Historical Society of Montana, with Its Transactions, Act of Incorporation, Constitution, Ordinances, Officers and Members* (Helena, MT: State Publishing Company, 1896): 50.

78. Isaac F. Marcossen, *Anaconda* (New York: Dodd, Mead, 1957).

79. Malone, *Battle for Butte*, 9.

80. LeCain, *Mass Destruction*; Fredric L. Quivik, "Smoke and Tailings: An Environmental History of Copper Smelting Technologies in Montana, 1880–1930," PhD diss., University of Pennsylvania, Philadelphia, 1998. One exception to this rule does exist. Jeffrey J. Safford, *The Mechanics of Optimism: Mining Companies, Technology, and the Hot Spring Gold Rush, Montana Territory, 1864–1868* (Boulder: University Press of Colorado, 2004), studies the patterns of rhetoric and failure in one mining district in early Montana, but he does not link its practices to the mining that came later.

81. Rickard, *A History*, esp. chapter 15, "The Gold, Silver, and Copper of Butte," 341–64.

82. On the theory of emergence as a function of complex systems, see Steven Johnson, *Emergence: The Connected Lives of Ants, Brains, Cities, and Software* (New York: Scribner, 2001). A succinct discussion of system dynamics can be found in David Waltner-Townes, James J. Kay, and Nina-Marie Lister, *The Ecosystem Approach: Complexity, Uncertainty, and Managing for Sustainability* (New York: Columbia University Press, 2008).

83. Rossiter W. Raymond, "Editorial," *American Mining Journal* 6 (July-December 1868): 1.

Montana is the richest mining country on the continent. It contains the silver of Nevada, with the gold of Colorado, with this important difference, that the mines of both classes are more extensive, and . . . far richer in Montana than in either of the localities mentioned. It is scarcely possible to conceive a finer field for legitimate speculation than this country now affords.

Hon. R. C. Ewing, "Report on Montana Territory" (1865)

Providence so ordains it that the superficial treasures of the earth designed to attract this enterprising class soon disappear, and a higher order of intelligence is required and a more permanent condition of things established.

J. Ross Browne, *Report on the Mineral Resources of the States and Territories West of the Rocky Mountains* (1868)

Next to the sailor, the miner has to withstand the most danger. The earth is against him, in holding with her rocky arm her treasures; so is the fire, which, in issuing forth from the depths of the mountain, is nourished by the air till it destroys in a blazing flame the timbering of the shaft; the water is his enemy in rushing from accidentally opened fissures, in roaring waves over his work, and so is the air, when it comes forth as fire-damps, or choke-damps, destroying the inhabitants of the subterraneous realms.

Adolph Ott, "On Mining" (1870)

TWO

The Value of Ores

Knowledge and Policy in Lode Mining Development

LODE OR HARD-ROCK MINING MARKED THE beginning of a kind of metal mining practice that had the look of an actual industry. It required a larger, more concentrated labor force; large, powerful machines; and large construction projects onsite. Hard-rock miners pursued gold and silver in underground tunnels, having blasted through solid rock in search of the treasure they hoped was embedded there. By its very nature, hard-rock mining required more brute force and sophisticated technologies than those used in placer gold mining; they were up against the energies of earth building. The rock could not

DOI: 10.5876/9781607322351.c02

be excavated with shovels, for example, but required picks and drills and even explosives. In addition, the metals had not been eroded out of their hard, rocky mineral veins like the placer deposits had, so hard-rock miners had to crush and sometimes burn and then crush the enriched metallic ores to break the metals free. For these reasons, the commitment to mine an underground mineral lode required patience and a willingness to commit time to development—the notorious "dead work" mining investors complained about. Hard-rock mining also consumed more energy and other natural resources and cost more money at the outset; it was more a middle-class than a poor man's pursuit.

Because hard-rock mining involved higher risks, required more intensive development, and pursued more uncertain ends, the absence of a federal mining policy left the western mineral lodes mostly unexplored until 1866. That year, as a result of concerted efforts by Nevada senator William Stewart, the US Congress passed the first mineral law in its history and promised to recognize "free mining" as a legitimate practice on all western public lands. Federal protection of what had been an elaborate and risky trespass onto public lands unleashed the pent-up energies that followed the end of the Civil War, and a hard-rock mining industry—at once wildly successful and wildly corrupt—began to grow around enriched ore lodes across the mountain west.

The first hard-rock mining was done to extract the highly enriched silver ores in the Comstock Lode in western Nevada and, a little later, ores in and around Leadville, Colorado. In both places, placer gold mining had drawn a mining population that came to know the broader mineral landscape and uncovered what would turn out to be valuable ore lodes. While most hard-rock miners in the 1860s and 1870s pursued the gold quartz in lodes that had partially eroded into the placer creek deposits, the enriched silver-sulfide ore lodes in Nevada and silver-lead-sulfide ores in Colorado proved that ore lodes held the possibility of creating unprecedented mining fortunes. Success with the Comstock silver ores created the first generation of western-made mining capitalists, who reinvested their mining fortunes back into the exploitation of potential silver and gold mines throughout the mountain west.

At the same time hard-rock mining burrowed deeper into the earth to extract ore and sent prospectors ranging the mountains in search of promising outcroppings, the federal government took an even closer look at its western lands—in particular its mineral resources—through the eyes of a new class of professionally trained mining engineers. This new attention raised serious questions about the first generation of western miners' ability to efficiently manage the complexities of lode mining and cast a shadow of doubt over the entire enterprise. Six years after the first law was passed, a second, revised federal mining law—designed to accommodate the spatial needs and property rights that followed the professional engineers' vision for scientific mining—superseded it. The result was a hybrid institutional landscape of open public lands and private techniques and technologies. The combination not only led to a mining industry quickly dominated by a wealthy elite, but it also set western mining development on a trajectory of iterative and seemingly inevitable institutional growth.

Despite its initiation on several gold and silver lodes in Montana concurrent with its gold rush and concerted efforts to organize mineral districts and launch lode mining after 1866, Montana's hard-rock mining industry failed to gain traction until the passage of the second federal mining law in 1872. The richest silver ore lodes in the state, located in a high mountain valley branching off of the southern Deer Lodge Valley, were composed of complex mineral combinations and required processing to extract their valuable silver. Bringing together local and regional capital with advanced training and experience for the first time in Montana, the Summit Valley and its small mining settlement of Butte suddenly emerged as the hard-rock mining center of the growing territory and a significant western silver producer. But as the hard-rock investors would quickly learn, keeping up with hard-rock mine investments was easier to imagine than to do.

MINERAL VEINS

Geologically speaking, mineral veins are secondary deposits of crystallized minerals within an existing primary rock formation. Veins result from hydrothermally induced geochemical reactions deep underground. They occur in subduction zones, where one plate of the earth's crust is pushed underneath another when they collide, creating magmatic activity and, usually, mountains. They are also formed at hot spots, like the region underneath Yellowstone National Park, where a particularly active region of the asthenosphere has led to an unusually thin crust. Veins are formed in these places during active orogenies, or mountain-building episodes, when molten rock rises high enough in the crust and groundwater penetrates deep enough in the bedrock to come into contact. The resulting hydrothermal energy, an explosive steam engine deep underground, forces the superheated solution into fractures and fissures made in the overlying rocks by the same mountain-buckling forces. As the solution—generally composed of silicates or sulfides and trace elements— moves toward the surface, falling pressure and falling temperatures cause the material to precipitate out of the solution, filling the open cavities in the rock. Taking the form of the crack or fracture in which it precipitated, the crystallization results in a rock formation with an organic name: a vein. The name derives from their likeness, where they have been exposed on the eroding outer surface of their country rock, to the narrow branching and winding channels of blood visible on the surface of

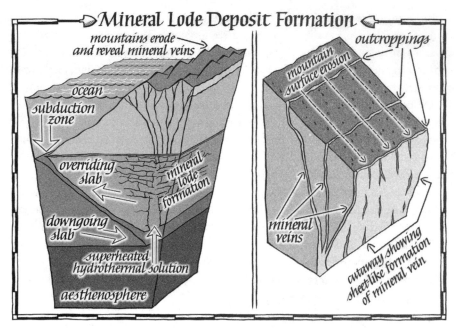

The formation of mineral veins or lodes involves hydrothermal deposition of mineral followed by upthrust and erosion of subterranean rock formations. The multiple fracturing of rocks during mountain-building episodes (orogenies) complicates veins' structure even more than these schematic diagrams reveal. Original art by Ruth Pettis.

an animal's skin. But unlike animal veins, which are essentially tubes that run along right under the surface, mineral veins' surface appearance is merely the narrow edge of a fissure extending deep into the earth.

These mineral deposits had weathered along with the mountain rock and, where they contained gold, could result in downstream placer deposits; but the outcroppings themselves also first interacted in place with rain and air and extreme temperatures, leading to chemical and mechanical changes in the exposed and near-surface material through oxidation and secondary precipitation. As a result, the condition of the vein at the surface was usually dissimilar to the condition of the vein farther down in the ground. For this reason, while mineral veins could be visible and obvious at the ground's surface, it was not wise to evaluate the constituents of any given mineral vein by the quality and contents of its surface ore; instead, it was advisable to dig deep enough to test unweathered vein material. The evaluation of an ore deposit always required an initial investment of time and money.

Mineral veins are ubiquitous throughout the mountain regions—their formation and deposit is a regular part of all mountain-building episodes—but few veins carry useful or precious metals, or the veins carry an insufficient content of the metal to be valuable. Mineral veins that do contain sufficient quantities of valuable metals are called *ore lodes*; after placer creeks, they became the next object of pursuit for western metal miners. The fact that a spatial relationship existed between valuable placer gold deposits and valuable ore lodes made this pursuit somewhat easier. Substantial placer deposits had resulted from the erosion of valuable ore lodes. For many of the placer deposits in the western mountains this erosion was still taking place, which meant that somewhere upstream and uphill from placer gold deposits, miners were likely to find gold-enriched ore lodes.

What was also true, although in no way understood by the throngs of placer miners who opened regions like Montana Territory to mineral exploitation, was that in places where valuable gold had been deposited, it was not unusual also to find a variety of other ore lodes containing gold, silver, and silver compounds as well as zinc, lead, copper, and iron. The metals are sorted out of the melted continent when the magma begins to cool because they are difficult to integrate into the rock matrix or crystallization pattern forming there. When metals are present in molten material, they tend to be chemically squeezed to the top of the magma chamber, where they are then more likely part of the solution first in contact with groundwater. If gold or silver had been present in the magma and been pushed to its rising face, some of it would have become part of the steam solution and been deposited in some way in the precipitated ore lode. In other words, while valuable mineral veins were, generally speaking, rare, when valuable ore deposits *were* found—in the case of the mountain west, by searching in the vicinity of a valuable placer deposit—it significantly increased the odds that additional lodes of valuable ore would be found in the vicinity; the "discovery" of placer gold often led to the discovery of gold and silver lodes.[1]

As much as gold and silver developed a particular kind of geological and geographic relationship during the formation and erosion of mountain ranges in the US West, they have also developed a cultural and social relationship in which their affinities and general similarities have provided a basis for differentiation and hierarchy. Both gold and silver are precious metals, materials too rare and too soft to have much utility for use as tools, and so they have often been used as adornment or currency—a cultural position that has led to the belief that the material itself possesses a natural value. But for all their similarities, gold and silver also ended up being fairly distinct. For one thing, silver is about seventy times more abundant than gold in the earth's crust. Its relative abundance, or gold's relative scarcity, causes it to have a relatively lower value than gold.

Even if silver is more rare, however, or if the rules of supply and demand are not immutable, it has another quality that assures a lower value than gold: silver is susceptible to the environment. Unlike gold, whose surface keeps its shine under all conditions, silver tarnishes in the open air. This happens both because of elemental silver's internal characteristics and because of the character of the earth's atmosphere—it is the product of physics, chemistry, and climate. The electrochemistry of elemental silver causes ions (positively charged molecules) to form on its surface. The silver ions, as charged particles, are like a new placer creek in a new mineral region in that they set in motion a rapid and cascading series of changes. In particular, they attract sulfur, which happens to be generally present in the ambient atmosphere; in bonding with sulfur, silver's surface tarnishes to a dull and ever-blackening color.

Silver's greater abundance and its affinity for sulfides means it can be more widely deposited in the landscape than gold. In addition to appearing in quartz deposits, silver is also often bound up in sulfide ore deposits, which are less common than quartz alone but are more often enriched with precious and useful metals where found. Thus placer gold often leads miners to gold quartz and silver quartz, which could then lead to ore deposits with multiple metal constituents in which silver is a likely component— such as galena (lead-sulfur), chalcocite (copper-sulfur), and enargite (arsenic-copper-sulfur)—as well as a wide variety of other mineral combinations.

Because of the geological and geographic association of gold and silver, the gold rush period not only mapped the US West as a mineralized region; it also—almost simultaneous with its earliest success—launched lode mining, also known as hardrock mining, by which process some western miners stopped digging up creek beds and began burrowing into mountains. By the early 1850s the California gold rush had begun to spill back over the Sierra Nevada, pushing what Rodman Paul has called a "mining frontier" eastward into the interior. Because of the age of the placer deposits in the Sierra foothills, a search for the source of the large creek deposits— the so-called mother lode—turned up nothing, so anxious gold seekers began to range eastward in search of new placer deposits. Within a year or two, hundreds of miners were working at productive placer gold locations across a region that would soon be as geographically segregated as Nevada, including 200 men at work in what they had named Gold Canyon. As these miners migrated up the creek in search of new diggings, they encountered a bluish stone they did not recognize. Subsequent assays revealed it to be rich silver sulfide, and by 1858 miners had found the source of the stones in several outcroppings of a silver-sulfide ore lode along the canyon walls upstream and above the creek. Word of these unusually rich silver deposits in an age when silver was an acceptable and used form of money set off a silver rush that drew tens of thousands of men from California and elsewhere into Virginia City, Nevada—the instant settlement that developed alongside these deposits where several of the earliest mining fortunes in the region would be made.[2]

The decision to pursue ore lodes, as the miners on Nevada's Comstock and subsequent other silver ore lodes would quickly discover, was a decision to engage in a different form of mining than the placer mining that had drawn so many people to the California foothills. First, the gold creeks offered almost no visual evidence that gold was likely to be buried beneath their detritus. In contrast, mineral veins often jutted out of the surrounding rock or otherwise distinguished themselves in an apparent or relatively easy-to-expose fashion.

But the obvious appearance of surface mineral veins was no guarantee that the ore contained anything of value. This was true even after placer gold had pointed the way to the location of potentially valuable ore lodes. The more visible quartz material required more intensive energy to work than the placer gold. Gold and silver quartz did not appear very different from one another or from quartz veins with no precious metal content; proximity to placer deposits was a helpful first indicator, but then the quartz—which was a rock formation—had to be extracted, crushed, and amalgamated (treated with mercury) to discover whether the ore contained precious metal and, if so, in what proportions. These difficulties grew exponentially with the wide variety of more complex compounds of sulfur, which often needed at least the additional step of roasting to release the silver or other metals from its bonds. Because silver and gold were rarely, if ever, evenly distributed throughout the ore lode, a sampling of ore that revealed a large amount of precious metals did not guarantee a regular supply of valuable ore as the lode descended into the earth. Thus it was generally wise to take several samples at different levels, which itself required more excavation. Conversely, the finding of no precious metals at the upper levels of the mineral vein did not preclude the existence of valuable ores below. In short, there were few rules of thumb in lode mining besides the need to work and sample.

However, in the wake of successful placer finds and especially after the production of silver from rich ore in the Comstock Lode created multiple mining fortunes in the late 1850s, the willingness to invest hope (in the form of labor and cash) into these surface appearances grew significantly. The hash marks William de Lacy traced throughout the mountains of the new Montana Territory in 1865 represented a clear expression of these growing hopes for the future of mining. Potential gold and silver ores were mapped and located on the ground all over the mountain west, no less in Montana, and they became the next objects of pursuit by western metal miners.

LODE MINING

Montana's lode mining industry, like Nevada's, began almost immediately alongside placer mining and then grew in intensity as the placer creek claims became increasingly worked over and a series of costly mines and mills went into production.

The first lode mining efforts were undertaken above Grasshopper Creek outside Bannack City during the winter of 1862–63. At that time a partnership of miners named Allen and Arnold constructed a water-powered six-stamp quartz mill out of wagon parts. They worked the soft red and brown oxidized quartz ores being mined out of two tunnels by a half-dozen laborers on the Dakota Lode 500 feet above Grasshopper Creek. When James Fisk visited Bannack in the fall of 1863, he reported that the Bannack quartz miners were getting upward of $2,000 in gold (about four pounds) out of every three-and-a-half tons of ore, although it took Allen and Arnold's handmade mill more than a week to process that much material. Fisk's report on the promise of lode mining included an important and revealing caveat: "Gold quartz mining requires capital and machinery and can only be profitably conducted by companies or associations." Only with some stretch of the imagination did the Dakota mine and the converted wagon stamp mill met these criteria, but Fisk was optimistic for the future. "Doubtless many will be operational next year," he wrote.

True to Fisk's predictions, several genuine prefabricated stamp mills were shipped to Montana in 1864, but despite John Mullan's work to build a road and the federal government's efforts to encourage steamboat traffic into the region, the isolation of the northern mountains made the heavy equipment rather costly—thirty-seven cents a pound on average from St. Louis or San Francisco—and put the necessary machinery mostly out of reach geographically. Even without stamp mills, however, prospectors continued to identify and claim ore lode deposits whose ore samples contained gold and silver in Beaverhead County north of Bannack, in Madison County in the mountains surrounding Alder Gulch, north of Helena in Lewis and Clark County, and in several locations in Deer Lodge County during 1863. All but one of these deposits was immediately in the vicinity of placer gold deposits, and all of them were quartz veins.[3]

The interest in trying to exploit ore lodes ranged beyond existing placer gold locations during 1864, when a few ambitious miners tried to reduce sulfide ores in several locations in the new Territory of Montana. At one location eighty miles north of Bannack City, across a high plateau, a company of German miners had organized a lode mining district and located claims on a three-foot-wide mineral vein containing what they believed was a sulfur-lead compound known as *galena*, which they hoped would be rich in silver. The men named the district Argenta (Spanish for silver) and systematically opened three shallow mines to access the ore, but the ore would not reduce with the mechanical means they employed (stamping). This fact damped the men's ambitions, and they did no more development in 1864. Elsewhere in Beaverhead County quartz prospecting continued, leading to the creation of the Hot Springs District, where the Helena claim—the district's single named ore lode—

appeared to contain workable gold quartz. Prospectors also created the Summit Mining District, where the Keystone claim appeared to contain valuable gold quartz. Without stamp mills, work had not yet begun in earnest in either location.

Two other lode mining districts in Lewis and Clark County, the Unionville and Ten Mile Districts, had also been named that year. The former contained potential silver ore and the latter quartz gold, but nothing had been done beyond naming the districts and evaluating the general character of the ore lodes. In the early years of the gold rush, placer creeks and placer mining took center stage in the region's mineral production. By comparison, during the same three summers in which prospectors and gold miners had identified the region's major creek deposits and excavated tens of millions of dollars in gold dust from thousands of individual placer claims, the region's few lode miners produced no more than a few thousand dollars in gold from four or five of eight named ore lodes. By the end of 1865, the year of peak production for Montana's placer gold miners, very little had changed for the lode miners other than the rise in the proportion of apparent silver to apparent gold lodes located.[4]

The high cost of freighting stamp mills and other capital equipment necessary for profitable lode mining was one obstacle to the development of hard-rock mining in Montana Territory in the mid-1860s, but it was not the only significant obstacle. At least as important, and with a negative impact on development, was the continuing uncertainty about the legal status of mining claims. It was one thing to work collectively with dozens or even hundreds of men in claims stretching for miles along a creek, as placer miners did. There was undoubtedly some comfort regarding the right to these grounds in the numbers present. Ore lodes, by contrast, generally required intensive rather than extensive development—a company of men digging vertically into the ground, usually working a single shaft. As the few developments concurrent with the gold rush show, lode mining districts were also often initially organized by a single interest trying to prove that the mineral vein was valuable and worth the effort and investment of others. This was a very different kind of attention to geology than placer gold received, putting more at stake in a single location and thus increasing the scale of individual risk. Throughout the 1860s—indeed, since the early 1850s—the US Congress had intermittently threatened to develop mining policy by which the nation-state would gain in the financial rewards of its mineral deposits through leasing, royalty, or taxation. Without a clear signal from Congress about the status of their find, about the security of land rights related to hard-rock mining, or whether the recovered metals would have to be split with the federal government, most men were unwilling to make much of an investment in the enterprise. Thus while a few districts were organized and a few lode claims were made, a lode mining industry never got started in the early years, despite ample signs throughout the region that there were valuable mineral lodes to be exploited.[5]

This uncertainty was finally addressed in 1866 with the passage of "An Act Granting the Right-of-Way to Ditch and Canal Owners through the Public Lands, and for Other Purposes," by which lode mining was first recognized as a federally sanctioned and protected activity within the public domain. The unusual name of the first federal mining law belied the unusual parliamentary techniques by which its champion, Nevada senator William Stewart, ferried it out of Congress, as well as the deeply contentious issue mineral policy had become in the wake of the Civil War for members of Congress representing the East and the Midwest. The bounty of gold that had flowed out of the western mountains and directly into the US economy had been taken from public lands and thus technically represented the theft of public resources. There was no precedent in the nation's history for such a vast and valuable exploitation of natural resources without at least a legal means of transfer and some form of agreed-upon method by which title to the land and resources could be regulated.

Congressmen and territorial representatives from the new mining regions in the mountain west, chief among them Senator Stewart, viewed the problem in a different light. They saw any effort to alter the system of so-called free mining as a threat to their survival because lode mining development was a promising foundation for the economic growth and settlement of their still sparsely populated regions. To this end, they sought a policy that gave the greatest incentive to lode miners. To pass such a law, however, Stewart had to literally sneak it past some of his eastern colleagues. He performed a legislative sleight of hand in which he amended a bill for rights-of-way on pubic lands that had already made its way out of committee with his own language for mining regulation. He then quickly shuffled his version of the bill through the final votes and to the desk of President Andrew Johnson, whose attention, like much of the rest of Washington's, was on the southern states in the wake of their Civil War defeat.

The 1866 law did not represent a consensus in Congress about mining policy; instead, it represented Stewart's best estimation about the needs of the still nascent western lode mining industry. The very specific and targeted goals of Stewart's legislation can be seen in that fact that it remained utterly silent on the issue of placer mining—the primary mining activity that had opened much of the mountain region and that, at least in 1866, did not seem to have run its course. More than likely, Stewart neglected placer gold because a mining policy that granted placer mining the same favorable terms offered to lode mining would have attracted more attention and perhaps foiled the effort altogether. It would be four more years before the mining law was amended to include placer claims, and by that time creek claims were offered at a quarter-section (160 acres), reflecting a scale that could only be worked through mass-production hydraulic mining.[6]

The 1866 mining law was designed to use federal policy and federal sanction to create the same energy around exploration, prospecting, and—where valuable minerals were found—extraction and milling that had animated gold rush activities since 1849. It did so by attempting to borrow from the spatial practices used in the organization of placer mining districts and adapting them to what Stewart understood to be the special needs of lode mining. The law created a new category of federal lands called *mineral lands*, a remarkably flexible and broad land designation that included any piece of public land on which "gold, silver, cinnabar, or copper" lodes existed. These lands were designated to be "free and open to exploration by all citizens of the United States." In places where lodes had already been located, the law recognized existing districts and existing district rules, but in all new locations lodes were to be divided into 200-foot claims for individual claimants and up to 3,000 feet for companies and associations of men, as long as at least $1,000 in labor had been invested to develop the mine. The cash value placed on labor investment suggested the subtle change in the social organization of mining accompanying lode mining, as placer mining districts had traditionally measured the investment in terms of days of work—$1,000 represented roughly the wages of six men over a seven-week period.[7]

Making a rough analogy with the flow of a creek channel, Stewart's law visualized the length of an ore outcropping to be analogous to the fall of a creek as it coursed across the land; the surface mining claims, like placer claims, cut perpendicularly across these outcroppings. The rectangular surface claim was designed to allocate adequate access to, and delineate the end lines of, the real object of interest, what the law called "veins or lodes of quartz or other rock in place." The property rights in the ore lode provided in the 1866 mining law, unlike any previous US property law, adhered to the ore lode itself and so had only a superficial relationship to the surface land claim. Placer gold had been deposited in a horizontal layer generally atop the hardpan of the creek, and thus the roughly rectangular claim on the surface of the ground could be projected perpendicularly down into the ground to the hardpan (theoretically, as with most common law, land property to the center of the earth) and provide a reasonable area in which to work gold. However, the quartz and sulfide ore lodes were not deposited horizontally; nor did they often descend vertically. The angle of descent, or "dip," of the plane of the ore body might, in fact, follow such an angle that it broke an imaginary vertical plane at the property line and entered into ground underneath a neighboring property. To contend with this fairly common occurrence, which had riddled the Nevada mines with conflict over ore extraction rights, Stewart's law granted miners the right to "follow such vein or lode with its dips, angles, and variations, to any depth, although it may enter the land adjoining, which land adjoining shall be sold subject to this condition." The lode was the object of possession; an invisible and unknown mineral structure had the power to supersede any prior visible surface claims that might impede a miner's access to his lode or challenge his right of exploitation. The 1866 mining law, Stewart hoped, would ignite a lode mining rush and send miners once again racing into the western mountains seeking gold and silver.[8]

To the lode miners and mine boosters in the western mineral regions, the new law provided the kind of certainty they were seeking. The *Sacramento Union* celebrated the fact that the bill had been "framed with a more intelligent regard for the interest of the people of the Pacific Coast than any other previous measure." Governor Richard C. McCormick of Arizona was pleased that the bill "preserves all that is best in the system created by miners themselves, and saves all vested rights under that system, while offering a permanent title to all who desire it, at a mere nominal cost."[9] Edmund Francis Dunne, a prominent Nevada attorney who ultimately felt the bill fell short in very important ways, first declared its positive effects: "This act at once restored order, certainty, and confidence, in [mineral] regions, by reason of the provisions of Section One, which recognized the binding force of local rules and customs . . . People knew where they stood; capital was invested with safety, and many districts sprang into new life, directly under the operation of this section of the law."[10] Lode miners across the mountain west had held their collective breath since the end of the Civil War, continuing to work at known locations like those in the vicinity of the Comstock Lode; the gold quartz near Central City, Colorado; and the Dakota claim above Grasshopper Creek; but few new claims were made as the miners waited to see what kind of policy would come out of Washington, DC. With Stewart's legislative success, however, the main cause of uncertainty—the right to prospect and possess ore lodes—was settled, existing claims were recognized, and new claims had a sanctioned procedure through which to become secure. Western lode miners let out a collective sigh of relief.

With the passage of the 1866 law, lode mining began to increase steadily in Montana Territory, and it began to expand beyond gold. In 1866, Montana miners located three new districts containing more than twenty individual mining claims. Only one claim out of two of the districts was presumed to contain gold; the rest were presumed to be silver lodes. The third district, located in Meagher County, had fifteen- to twenty-foot-wide mineral vein outcroppings presumed to contain copper. The following year, five new lode mining districts were named in the territory, again predominantly around silver lodes, adding another half-dozen or so individual lode claims to the territory's total. The technological infrastructure needed to support these growing numbers of lode claims grew in-kind. Only two stamp mills were found in the region in 1865; that number grew to seven by 1868 and to twenty-one in 1869. Stamp mills worked quartz gold and free-milling silver, both of which could be pulverized and washed with mercury to extract the precious metals. By the

*Five-stamp mill, the basic production equipment during the quartz mining period.
These mills were heavy and expensive to transport, so investors tried to encourage quartz
exploration whenever they brought such equipment into the territory. Courtesy, Wiki
Commons.*

summer of 1871, thirty-seven stamp mills were dropping more than 500 stamps on gold and silver quartz in the twenty-one lode mining districts scattered around the five southwestern counties of Montana Territory. By this point, gold had also ceased to be the primary mineral sought by mining developers, as more and more mineral claims were composed entirely of silver ore.[11]

It had become obvious in the years following the Civil War that silver deposits were much more abundant than gold deposits in Montana Territory, but creating lode districts and locating claims was the easy part of silver mining. Figuring out how to extract the value from the ore in a silver mine was another problem altogether, especially if the claim held refractory, or difficult-to-reduce, ore. By 1866, Samuel T. Hauser had begun to accumulate significant cash reserves. Having arrived with the original Fisk Train in 1862, survived Crow attacks at Big Horn River on the Yellowstone Expedition, and joined the vigilantes in executing reputed members of the Plummer highway gang, Hauser then made a fortune in real estate as an early property owner in Helena. When the US Congress sanctioned lode mining on public land in 1866, Hauser decided to try his hand. He acquired the right to and began working the silver mines that had been opened in Argenta a few years earlier. He partnered with two businessmen from St. Louis and hired two German-born miners, Augustus Steitz and Philip Deidesheimer, to build a small German-style lead-smelting furnace near the mines. Deidesheimer was well-known for inventing square-set timbering on the Comstock Lode, an innovation that had saved silver mining in that district, but he could not successfully reduce the Argenta ore in his furnace, for reasons he did not entirely understand.

In 1867 Hauser gave up on this venture and sent the same men to scout the ore lodes in the Flint Creek District, where several claims had been located on what appeared to be silver deposits. The men reported that the ores in that location resembled ores they had seen at the Comstock, so Hauser directed them to construct a mill similar to those used on the Comstock ores, using a process called *pan amalgamation*. In this process, the pan was heated by steam hot enough to evaporate the sulfur in the ore and precipitate the silver into a mercury flux that was added to the heated ore. Hauser hired James Stuart to superintend this mill—the James Stuart Mill, named for the leader of the fated Yellowstone Expedition—and he platted a settlement near the mines that he named Philipsburg, for Philip Deidesheimer. The promising development and new settlement soon drew several hundred optimistic miners.[12]

Montana lode mining, like lode mining throughout the western mining regions, had begun alongside the activities of a gold rush; because of the region's extensive mineralization, however, it soon took on a new geography all its own. Indeed, in the years following the peak of placer gold production in 1865, the continued flow of interested miners and heavy freight into the region—partly a response to the steadily growing production of gold from the region since 1862 and partly the result of pent-up energies that, until 1865, had been focused on the Civil War—provided both the labor and the fixed capital needed for the development of a lode mining industry. The successful passage of the 1866 mining law provided federal sanction and policy certainty regarding the disposition of existing ore lode claims and the means to legally secure future claims, also necessary to provide enough confidence for the level of investment needed to prove up a lode mining claim. In general, the law further bolstered a favorable environment for lode mining development. Together, these external forces pressed the continued development of lode mining in Montana—first in search of additional gold quartz deposits and then of silver and copper deposits, which, as Hauser's experience shows, required both an additional level of investment in a mill more specialized than a simple stamp mill and new levels of expertise to develop the right processes by which to work the more complex sulfide ores. As the 1860s wound to a close and new silver districts and mills began to dot the territory, the emergent lode mining industry seemed to be compensating for the decline in placer yields in the region.

THE FREIBERG MEN

One small but very important section of Stewart's 1866 mining law granted the president of the United States the right to "establish additional land districts and to appoint necessary officers" to fulfill the provisions of the law. Mineral policy was ultimately about the production of specie (gold and silver), so President Johnson assigned a commissioner of mining statistics to the US Department of Treasury under the charge of Treasury secretary Hugh McCullough. The first appointee was a western journalist and novelist, J. Ross Browne, who had been in California since 1849 and played several key roles in its establishment as a state and in federal investigations of misconduct in the region. Browne, however, was neither a mining expert nor willing to travel very far from his home in San Francisco, instead communicating with mine owners and developers in the various regions through correspondence. The resulting reports were engaging but tended to be overly general and inconsistently accurate. While Treasury and Congress wanted an assessment of the western mining industry, it was clear that western lode developers and regional observers saw the reports as an opportunity for free publicity by which to promote districts and claims that had yet to prove their value. In early 1868, Browne was replaced by eastern publisher and mining engineer Rossiter Raymond, a decision that would mark the beginning of a new era for western metal mining.

During the 1850s and 1860s, while gold seekers chased rumors and washed potential pay dirt in placer gold diggings throughout the mountain west—in effect mapping much of the region as a potentially valuable mineral landscape and creating new patterns of natural resource exploitation—a few dozen men from upper–middle-class families in the East took off in a different direction, seeking a different kind of knowledge about minerals. Instead of heading west in search of fortunes, these young men spent fortunes and traveled east across the Atlantic Ocean in search of scientific and technical knowledge about mining. They sought a body of skills and understandings produced by the mining practices developed in and around Freiberg, Germany, which were offered for sale through the Bergakademie, an institution of higher education organized around mining. Protestant, white, generally Republican if not abolitionist in their politics, and often reformist in vision, men such as Rossiter Raymond, James Hague, Arnold Hague, W. W. Keyes, Edward Peters, Richard Rothwell, Eckley Coxe, Anton Eilers, and many others spent two to three years studying what they came to call mining engineering in one of the oldest and most developed mining regions in Europe, where geology, chemistry, and mining engineering had become formalized around the production of metals over more than a century.[13]

The US students in Germany had been learning about a nature that, because of the region's history, had become much less uncertain than the one encountered by western mining pioneers. For one thing, after centuries of development, European minerals and mineral deposits were well understood; both the inorganic nature of chemical bonds on the micro-scale and geomorphic shifts in ore deposition on the macro-scale had become relatively legible through long-term mining efforts. The German mines and mills were used as object lessons for the American students, indelible examples of the great potential of the systematic and scientific development of mineral resources. Equally important, the region of Germany in which these men studied mining had long exhausted its high-grade ore lodes, presenting a geological history and a set of geological conditions that provided a twofold benefit to the Americans. First, it showed them the value of careful development, suggesting that even where rich ore lodes were evident, such richness would not last forever; for that reason, it was imperative to develop mines carefully.

Perhaps even more important, the European developments and efforts by the Bergakademie to expose its students to both theory (chemistry, geology, and techniques) and practice (actual visits inside mines and participation in smelting processes) helped them understand and eventually visualize for the future US mining industry the great possibilities presented by well-developed deep mines and the much more abundant low-grade ore often found at depths. The mining region of north-central and eastern Europe had been mining its ores for a long time; the combination of a smaller population and less significant market demand, combined with the fundamental need to develop needed techniques as they went, caused the mines to be developed slowly and systematically and the processing facilities to be gradually adapted as changes in ore required alterations to the technique.

This single mining center in Saxony, where the mining school was located, had developed silver- and lead-smelting techniques—allowing the students to work with ores from its own mines, as well as with various ores purchased from elsewhere in mining regions around the world. The three different types of smelters together had the capacity to work several hundred tons of ore material each week. The mining center's mills produced more lead and silver than anyplace else in the world. Students at the Bergakademie were taught assaying methods to identify the constituents of ores, as well as various techniques developed to reduce the ore and extract the desired elements. These methods and techniques had been designed through craft practices separate from the European scientific revolution, but by the mid-nineteenth century the language and practices of relevant scientific disciplines had become critical tools of industrial low-grade miners and provided a common vocabulary by which German miners, among others, could communicate their techniques to others.

As a result, students at the Bergakademie were taught principles of geology, geomorphology, and geognosy inasmuch as these sciences informed the exploitation and working of ore lodes. They also learned the well-honed practices of subsurface surveying and mine management, critical skills for effective lode mining. Students visited mines across northern Europe, examined ores, and learned the chemistry of existing smelting techniques. They were exposed to the vast range of practices representing the height of nineteenth-century mining knowledge and practice. The most ambitious of these students, usually from the United States, took advantage of their location and ventured far into Silesia (present-day Poland) to examine coal mining practices. Most of them also made a pilgrimage to the southwestern shores of Wales, where miners and smelterers had developed the world's most advanced and productive copper smelters. In the industrial port city of Swansea, Welsh smelter men had learned how to reduce the wide variety of copper-bearing ores found in the Cornish countryside and by mid-century had become the world center of copper smelting and production. Through their European education and their exploration of these mining and smelting locations, American students were shown that long-term mining was possible.[14]

Each of the European mining centers had developed extensive and sophisticated production systems that depended on the accurate identification and efficient mass movement of ore resources. European miners had learned to open mines systematically and carefully, identifying ores and calculating reserves as they went. They obsessively sampled, sorted, mapped, and recorded. Smelter men tracked the flow of minerals with visual skills that were being reinforced by modern chemistry, giving the impression that nothing in the process escaped oversight. In this way, by the mid-nineteenth century Germany and Great Britain had emerged as the most significant and important metal production regions in the world. Learning their trade in these places gave students from the United States tremendous confidence in their ability to understand the needs of mineral production when they returned home, as well as a toolbox of skills and techniques to draw from as they put their education to work.[15]

Initially, many of the Freiberg graduates went to work wherever they could find a need for their services, which limited them to the East, upper Michigan, and eastern Colorado. In these locations lode mining had begun, there was enough capital investment to afford consultants, and there was easy access by railroad. James Hague first went to the Lake Superior copper mines after graduating in 1858, and he used his new knowledge to help copper producers there improve the efficiency of their extraction and reduction processes. Anton Eilers, who graduated with Rossiter Raymond in 1860, traveled back and forth between Colorado and New York in an effort to impose some order on what had become a frenzied speculation crisis among mining properties. After two-and-a-half years in the Union Army, Rossiter Raymond joined forces with Julius Adelberg, a New York–based mine industry consultant with clients in New England, Pennsylvania, and the mid-Atlantic states. In this fashion the Freiberg graduates began to acquire a broad-based knowledge of the US mining industry at mid-century. They evaluated mines and mills up and down the eastern seaboard and as far west as Colorado. They began to produce meticulous studies and conservative evaluations of the mining operations they examined. Because of their experience and training, they tended to view American mining developments through the eyes of the long-established European examples from which they had learned their profession. The result was consistent disappointment in what they found.[16]

For all of its energy and ambitions, they discovered, US mining was generally unprofessional, inefficient, and ultimately wasteful in its practices. But "all difficulties relating to the management of ores and the extraction of the gold will vanish before advancing science and engineering," Raymond asserted in an early report. "The mine must be put in charge of an Engineer capable of forming a plan and adhering to it," he wrote in another. The men worried about the tendency of inexperienced miners to quickly extract the richest ores, often undermining the true value of mineral deposits in the process.[17]

By the mid-1860s and in the wake of Stewart's national mining bill, Raymond made several efforts to expand his influence, and that of other European-trained mining engineers, by creating more formalized institutions to promote and share their work. In 1866 Raymond and Adelberg organized a mining consulting firm, to which they gave the very official-sounding name the American Bureau of Mines, "for the purpose of assisting to place the mining enterprise of this country on a sure and conservative basis"—conservative, in this case, in the same sense of the word adopted by the forest conservation movement later in the century. Public resources such as mineral ores, according to their logic, should not be wasted, and leaving these resources in the hands of untrained gold seekers or men with no experience in mining could lead to no other result. Raymond and Adelberg's bureau offered certification of engineers to help mine owners distinguish legitimate engineers from charlatans and quacks. James Hague, among many others, subscribed to this effort. In 1867 Raymond purchased and began editing the *American Mining Journal*, which he renamed the *Engineering and Mining Journal* in 1869 "to express more accurately the comprehensive character which we have attempted to impress upon the *American Mining Journal*." By the 1870s the publication had become the central voice of professional mining in the United States, a status it still maintains today.[18]

As the Union Pacific Railroad began to gain on its connection with the Central Pacific Railroad in Utah in the late 1860s, several of the Freiberg men were chosen to contribute to what was becoming a growing federal interest in western mineral lands. Clarence King, the future creator of the US Geological Survey, tapped James Hague to serve as chief geologist on King's survey of the 40th parallel, a region within striking distance of the Union Pacific. Hague was to offer a description and assessment of the mining developments and mineral potential in and around the Comstock Lode in Nevada and the Leadville mines in Colorado. In late 1867, after two disappointing years with J. Ross Browne in the position, Treasury secretary Hugh McCullough appointed Rossiter Raymond to serve as commissioner of mining statistics. Raymond, in turn, appointed his Freiberg classmate Anton Eilers as his deputy and enlisted the expertise of several other Freiberg graduates who had moved to various locations in the western mountains. Raymond's job was to solve the problem, as Treasury saw it, of a declining gold yield in the western mining districts. During the summer of 1868, Hague, Raymond, Eilers, and a handful of others brought their European training on behalf of the US government to the nation's western mining districts, and the US mining industries would never be the same.[19]

King and Hague eventually produced *Mining Industry*, volume 3 of the King survey, a monumental study of western economic geology. This 600-page tome explored the geological formation of Nevada's Comstock ores and Colorado's Leadville ores and provided a carefully detailed description of the mining businesses and developments in operation in both districts. Echoing the critique already emerging in consulting reports, James Hague found that even in Nevada's advanced fields, there were mining expenditures that could be avoided under more comprehensive or consolidated management. He also complained about mining and processing techniques whose efficiency could easily be improved. King's geology and Hague's mining critique stood as shining examples of the kind of professional and scientific attention

America's mineral districts could receive and heralded a new age in both visual and textual mining representation in the United States, but Hague returned to private consulting after the survey was complete.[20]

Rossiter Raymond was another story; his background, growing up in Henry Ward Beecher's abolitionist and reformist Plymouth Church in Brooklyn, no doubt fueled the scope and character of his ambitions. After his studies in Germany, however, Raymond wanted to bring the benefits he saw in the European mining industry to the United States. After a few years of coming to understand the character of the US mining industry, he became especially zealous about the cause. The state-supported social reproduction of mining knowledge linked the productive power of the mines and mills to the social reproduction generated at the university in Saxony. Raymond came to believe the same practice could work in the western United States, and he used his report to the US Congress in 1869 to make this argument. Raymond's report on western mining conditions, which was almost 500 pages, represented the most comprehensive survey of the developments and mineral geography of mining in the US western mountains compiled to that point. It treated each western mineral region in careful detail, listing districts and mines and ores and yields for every location willing to provide information to Raymond or his many agents and contacts. Raymond assembled a mountain of data, describing ore bodies and mines, explaining methods of extraction and methods of ore treatment. The report also listed and compiled volumes of ore removed, estimated and known production values for mines and mills, and assessments of future productivity.[21]

Western miners may have hoped that so much careful attention from Treasury officials would help to attract capital and businessmen to their regions and contribute needed investment to the lode mining promise of so many districts, but Raymond's report was anything but a simple and glowing promotion of the western industry. Indeed, it was obvious from the tone of the report that Raymond found much to be desired about mining developments in the US West. Raymond had little patience for what he considered the trial-and-error methods of the placer gold and early lode miners.

By holding the young western mining industry up to the impossible standards of long-established European practices, Raymond articulated a concern that things had started badly and that the consequences were increasingly dire if the industry continued along the same trajectory. He conceived of the minerals buried in ore lodes to be the property of US citizens, public property and a national endowment, and he believed the US Congress ought to do everything in its legislative power to ensure that this public endowment be treated with care. Raymond reminded Congress in his first report on the West that wasted and lost minerals were gone forever. Knowledge of these kinds of limits had generated an ethic of efficiency among

Rossiter Worthington Raymond (undated). No single individual did more to popularize and promote the interests and perspectives of professional mining engineers during the last third of the nineteenth century than Raymond, a New Yorker trained at the Freiberg School of Mines. He did not succeed in realizing his full vision, but he did create the foundation for professionalization of the field of mining and metallurgy in the United States. Courtesy, Wiki Commons.

German and Welsh miners—every possible bit of a resource ought to be recovered from start to finish. European developments had built this ethic into the tools and techniques they used. Down in the mines, mine foremen and managers developed tunnels along ore faces and sampled carefully to develop as accurate a sense of ore reserves as possible. Aboveground, mill managers could enlist the technologies of knowledge borrowed from chemistry and the ability to generate and control incredibly high temperatures inherited from alchemy to work toward better and better achievement of complete material recovery. For the US men who had seen and learned about the practices and skills needed to accomplish careful mineral extraction, developments in the US West created an anxious fear of impending limits and soon-to-be lost opportunities.

"Mining has been found in too many instances to be unprofitable," Raymond reported in the introduction to the first report he made to Congress, because it had developed "in a lawless and careless way, without much regard for the future." Raymond suggested that the problems facing western mines (and, indeed, mining in general in the United States in 1868) were not typical of business difficulties in other industries because mineral resources were an elusive rather than a fixed target. The days of frenzied gold rushes were coming to an end, Raymond pointed out, but the shift to new sources of ore—particularly the exploitation of quartz lodes and sulfide ores—required a new set of skills and competencies that had not been developed within the culture of placer gold mining.[22]

For Raymond, the rapidly falling production of placer gold was not about the limits of nature but instead stood as material evidence of a failed strategy. Placer gold production should not have declined so quickly and had done so only because of poor planning and an even worse execution of work. The crude methods and frenzied pace with which placer miners had worked their claims had left "waste" piles littering placer creeks that continued to hold significant gold, but that gold had been diluted into the detritus to such a degree that it could no longer be economically separated. An absence of planning had also caused other valuable mineral deposits to lie buried under piles of waste or to be overlooked entirely.[23]

The consequences of these practices were not only the unnecessary waste of the nation's mineral endowment but a concurrent and significant misallocation of capital resources as well. In some cases, uneducated owners of mineral claims either misunderstood or misrepresented the value of their claims, drawing investors to unproductive mines. In other cases, in their rush to access underground ores, miners had developed poorly designed mines and tunnels that had permanently buried otherwise good ore. All of these activities led to careless speculation and a waste of capital as significant and permanent as the waste of valuable natural resources. Even the 1866 federal mining law, which had made many western miners happy by sanctioning their practices and giving them clear title to their ore lodes, contributed to this situation by legislating uncertain and confusing ownership rights. The law's provision that each mining claim had to be limited to one stretch of a single outcropping had caused otherwise productive mines to be unnecessarily tied up in litigation as owners made competing claims on the same underground ore lodes, wasting even more valuable capital on legal expenses.[24]

The wasteful and inefficient practices of gold mining pointed to even more troubling future difficulties just around the corner for western miners. As these miners had moved out of the exhausted gold gulches into ore lodes, they created two new sets of challenges they had already proven themselves unprepared to face. The continually expanding extraction of ore lodes necessitated the excavation of mines deeper than any previously sunk in the region, Raymond warned, and deep mines required a battery of skills that were simply not present in the western mining culture. Equally worrisome to Raymond was the fact that enriched quartz ores would undoubtedly be exhausted in the not-too-distant future; if mining hoped to remain successful in the region, it would require the exploitation of sulfides. Unfortunately, however, sulfide ores were also much more complicated and expensive to work, requiring technical expertise and advanced knowledge to identify the ores and much larger investments of capital to afford construction of proper reduction techniques. An instinctive reformer, Raymond believed action could be taken to change the industry's course and bring about an alternative future: "When the industry of mining in these rich fields is based upon a foundation of universal law, and shaped by the hand of educated skill, we may expect it to become a stately and enduring edifice, not a mere tent, pitched to-day and folded to-morrow."[25]

The problem stemmed, Raymond believed, from American miners' failure to fully grasp the true nature of mineral deposits, and perhaps the pouring out of gold from the western regions also gave the US Congress the wrong impression about its public mineral holdings. The advantages of having great mineral wealth, Raymond warned, were "counter-balanced by the fact that its sources are not perpetual." He lamented that the minerals thus far wasted in the western mountains were "irretrievably" lost.[26] The longer the United States allowed its resources to be exploited by untrained men with inadequate technologies, the greater the volume of American mineral potential that would be lost forever. "Mineral wealth was given to a country but once," Raymond concluded. "Such resources would not grow again, like products of the soil."[27]

The first half of Raymond's Treasury report on the western mines provided a detailed description, organized by state and territory, of the entire western mining industry—describing mines, mills, and ore lodes and identifying shortcomings in knowledge and technology. The second half of the report offered policy recommendations. Raymond warned Congress that mining had not received adequate and necessary attention from the United States. He paraphrased Agricola to emphasize the industry's importance: "Mining and agriculture are the two great forms of productive industry. Strictly speaking, agriculture is the most important, since without it men could not exist; yet mining is almost as essential, since without it men could exist only as savages." Raymond believed the most important steps needed to place the industry on firm, permanent footing in service to the growing nation would be to create a national mining school somewhere in the western mountains and to develop regional expertise through the practice of scientific mining on western ores. A federal investment in the social reproduction of mining knowledge based on western ore production and the institutional support of scientific investigation and study would provide the strongest foundation for the efficient and conservative exploitation of mineral resources and lead to a robust and permanent US mining industry.[28]

Beyond providing the necessary training, Raymond also suggested that the federal government rewrite its mining law in a manner that borrowed from the best practices and policies of the German system while simultaneously preserving the essential political and economic characteristics of the United States. By this he meant secure and unambiguous title to ore lodes and surface locations combined with a public commitment to the sharing and exchange of mining knowledge and techniques: "The true and only course which the federal government can take in this discharge of its duty as trustee of our vast mineral resources is to establish some agency for the spread of knowledge among the mining communities, and to facilitate, in every possible way, the development of those communities into regular forms of society." For Raymond, this meant the creation of a national bureau of mines that would gather and maintain statistics and disseminate information about technologies and ore, as well as technical schools to train miners. These developments were of critical importance, Raymond reaffirmed, if the United States wanted to efficiently exploit its remaining wealth in quartz lodes and compound sulfide ores.[29]

Raymond's recommendations for comprehensive mineral policy reform fell mostly on deaf ears. Senator Stewart put forward several mining school bills, each more tepid than the last, but each of the bills failed. The challenges of Reconstruction and postwar national governance took priority over Raymond's ambition that Congress launch what amounted to an entirely new kind of natural resource agency and the establishment of educational practices the federal government had never before made its own. So Raymond took matters into his own hands. In 1870, the year after the publication of his first report, Raymond joined with Eckley Coxe and Richard Rothwell—fellow Freiberg graduates and experts in eastern coal and iron mining—to create the American Institute of Mining Engineers (AIME), an exclusive forum in which to share professional knowledge and to confront problems and develop collective solutions. That same year, Raymond founded the *Engineering and Mining Journal*, the "official organ" of AIME that published notices of meetings, meeting minutes, and featured papers. Shortly thereafter, Raymond's Scientific Publishing Company began publishing editions of his US Department of Treasury reports on the western mines under its own title—often well ahead of the congressional copy—as well as *Transactions of the American Institute of Mining Engineers*, which contained select papers and minutes from AIME's quarterly meetings for distribution to all members throughout US mining districts. In this way, Raymond positioned himself at the center of an emergent professional US mining culture, one for which he had created the textual and institutional presence out of whole cloth. More than any other individual, Rossiter Raymond would come to symbolize the first generation of professional mining engineers in the United States, and he would remain among its most prolific writers and influential champions well into the twentieth century.[30]

TREASURY VISIT

Raymond's work for the US Department of Treasury did not go unnoticed by boosters and developers around the mining west, not the least interested parties in Montana Territory. When Raymond's first report reached the desk of James Miller, the editor and publisher of *New North-West*, a territorial weekly newspaper published in Deer Lodge City, Montana, Miller was irritated. Two summers had come and gone since Rossiter Raymond had taken over the position of mining statistics commissioner, Miller complained in a pointed editorial about the mining report, and Raymond had already published one report and finished gathering data for a second report, yet he had failed to show up in Montana in person. Miller was particularly incensed because, even in 1869, Montana Territory was still second only to California in annual gold production, a status that all but obligated Raymond to visit personally. Miller was also keenly aware that placer gold production had been in a steady decline since 1865, and he knew as well as anyone that the territory's future mining hopes rested with lode mining, which he believed was being held back only by the slow investment of capital in the region. Seemingly oblivious to the irony of a region that had produced $100 million in gold crying out for lack of capital, Miller lashed out at Raymond, accusing him of neglect of duty and waste of public money.[31]

While Miller wrote from a comfortable office in Deer Lodge City, a bustling rural trading center nestled amid rolling grassland where John Grant's cattle still grazed by the thousands, he was every bit the champion of mining. He believed the purpose of Treasury's work in the mining west was to help mining regions make their resources known, to connect mine owners with interested eastern capitalists and attract investment. He also undoubtedly believed he was writing his critique essentially for the eyes and sense of righteousness of his Montana readers. In other words, the larger goal of the editorial was to sell newspapers to Montana mining men by sympathizing with their plight. As he would soon learn, Miller was wrong about Raymond's interpretation of the purposes of the editorial, and he had miscalculated its impact.

Within weeks of Miller's attack on Raymond, the federal commissioner responded, using his own national *Engineering and Mining Journal* to publish a lengthy defense of his work for Treasury and to launch vitriolic counterattacks against Montana's mine owners. Raymond explained that his meager budget each year did not allow him to travel to every mining region in the West or to hire more than one direct assistant. He expressed personal pride in his ability to have covered as much territory as he had since becoming commissioner, going to different places each of the first two summers, as well as in having found inexpensive strategies to acquire information and data from regions he had neither the time nor the resources to have yet visited. In these cases, he had enlisted fellow Freiberg gradates who had relocated to various regions in the West to survey mine owners in their region and provide him with reports. In Montana, for example, he had depended on W. W. Keyes, who lived in the territory and in fact had authored most of the material on Montana in the first two Treasury reports. Raymond complained that Montana's mining men had done very little to help their own cause when they neglected to provide much of the information Raymond had asked Keyes to collect. While Raymond adopted a stern and dismissive tone toward Montana miners, the fact that he responded swiftly and publicly reveals the ultimate importance of Montana's mineral production for his vision of the mining industry: he knew a region that had produced so much gold was bound to be enriched in multiple ways. Less than two years later, in the summer of 1871, Raymond and his deputy, German-born and Freiberg-trained mining engineer Anton Eilers, rode a stagecoach 400 miles from Corinne, Utah, to Virginia City, Montana, to see for themselves.[32]

The Montana mining region comprised 25,000 square miles of rugged mountains and well-drained valleys, and while the promise of its mineral resources had been made legible by the labor of Montana's placer gold miners and a growing lode mining industry, that labor could not make the region smaller. When Raymond and Eilers looked at a map of Montana, they determined that the best approach would be to travel separately, dividing the territory in two along a north-south axis. They split up in Helena, and each toured his part of the region over a three-week period during late July and early August 1871—just nine years after the first waves of the gold rush had begun arriving in the region and five years into the lode mining development that had begun following the 1866 federal mining law.

Raymond's relationship with Montana had not started in a genial fashion in 1869; and the 400-mile, dusty, bumpy stagecoach ride necessary to arrive at the mining centers in 1871 did not improve his mood toward the place. Within days of their arrival, Raymond and Eilers began publishing reports from Montana Territory in the *Engineering and Mining Journal*. For all of its accomplishments during the gold rush era and the many developments around mineral lodes, Montana did not receive favorable marks from the professional engineers. Raymond began by complaining about the region's isolation, based on which, he wrote, "it seems marvellous that anyone should come here at all."[33] In subsequent weeks, Eilers added the critique that those who *did* come to Montana were entirely ill-prepared for what they found. Where the *New North-West* and Montana's miners believed they had created tremendous promise for the region, Raymond and Eilers found waste. They described gulches in Bannack, Virginia City, and elsewhere in Montana Territory that had been worked only for the easiest gold and then left practically idle. Raymond characterized mining labor in the territory as demoralized and misguided in its agitation against Chinese immigrants to keep wages artificially high and hold unworked claims for better days. He described placer claims where the cost of extraction amounted to almost 80 percent of the value extracted and local markets that allowed poor miners "to go into a tacit, informal sort of bankruptcy" because of a "public sentiment which disapproves [of] 'being hard on a man when he is down.'" "This sort of generous feeling in the community is pleasant to contemplate," Raymond wrote with some contempt, "but it is demoralizing to business, and can only exist when mining is, on the whole, prosperous, times are 'flush,' and merchants are able to secure large profits, to cover their large risks and losses." These conditions did not then exist in Montana's placer mining centers.[34]

Montana's promoters might have pointed away from the placer creeks to the ore lode development work that had already begun—the stamp mills pulverizing tons of ore every week, the smelters working to reduce silver and lead, the hundreds of miles of ditches etching the landscape supplying the needs of hydraulic mining.[35] The territory had produced an average of $12 million a year in gold since 1863, which still ranked it second among states and territories in annual gold production, but the engineers would concede nothing to the region. "The mills of Bannack are nearly all old and imperfect and sadly out of repair," Eilers sneered. He then rode north to Argenta, where he found Hauser's furnace and two copycat furnaces: "In the neighborhood of Bannack are 3 smelting works, or what appear to have been intended such. Two of them, into which I had access, bear evidence that the builders had not the slightest idea about metallurgical operations, the third was closed up." Hauser believed that, with another season of work, he and his engineers could perfect the process they needed to reduce the Argenta ores to silver. Eilers was not optimistic; he suggested that Hauser's mining experts had failed to accurately identify the ores. "All of [the smelters] are of course idle," Eilers wrote, "there being no lead ores in the vicinity, and even if they were present, such works could never be conducted profitably."[36] In a longer piece about Montana's silver industry, published in the *Engineering and Mining Journal* in the fall of 1871, Eilers generalized the situation he found in Argenta as symbolic of a larger problem across the US western mineral regions: "True to the usual mode of developing mining districts in the West, several parties rushed to the conclusion that the ores of these mines must be smelted, though there was no mine opened to a greater depth than twenty-five feet."[37]

Raymond, who had a chance to visit Hauser's other development in Philipsburg—which by 1871 was also abandoned—offered a similar critique of the skill and talent in the western part of the region: "Most of the metallurgical operations in this direction in Montana have been unsuccessful." The reason, Raymond explained, was the absence of trained mining talent in a region just beginning to encounter and contend with complex ores, combinations of mineralized material that required more than just gravity for processing. "There is a dearth of skilled workmen. The millmen of Reese River, Cerro Gordo, Georgetown, Eureka, and other districts where refractory ores are either amalgamated or smelted, do not seem to wander hither. I saw ore shipped to San Francisco the other day, which could be treated on the spot, if the man who had tried it and failed had only known how to build and run a common Reese River reverberatory—to say nothing of such improved apparatus as that of Stetfeldt or Brückner," Raymond complained, listing patented smelting processes about which most Montana miners had never heard, never mind had any idea how to construct.[38]

Raymond's and Eiler's critiques of Montana mining were reprinted almost verbatim, with additional conclusions, in the final federal report on western mines for the US Department of Treasury for 1871. "In isolated districts like those of Montana," Raymond wrote, "there is frequently a surprising ignorance of what has been done elsewhere in the way of determining the best machinery and processes and the miner frequently wastes his time and money in experiments which have long ago been rendered unnecessary." Raymond cited the high costs of transportation and labor and the continued lure of creek gold in the warm seasons as contributing factors. He also blamed the region's shortcomings on the characteristics of the gold miners who had peopled the territory: "Alluvial mining attracts a population usually without capital and not specially experienced in the very different requirements and risks of quartz mining. Not realizing the complex nature and amount of the expenses attending the extraction and reduction of ores, the gulch miner is apt to be overly sanguine in his estimate of the value of veins, and to underrate the difficulty of working them to permanent profit."[39] Raymond also worried about the driving interest from outside investors who sent money into the region without hiring competent local overseers: "There has been in Montana at least the usual proportion of wild investment by eastern capitalists, reckless mismanagement by incompetent or dishonest agents and plundering by everybody of the non-resident owners who seem to be considered fair game in many mining districts."[40] The conclusions were harsh, and the picture was negative. By the time he finished reading Raymond's reports in 1871, James Miller may have wished that the commissioner had continued to avoid Montana.

While Miller and others in Montana could not have helped but take Raymond's and Eiler's critiques of their region personally, the Freiberg's men's assessment, in fact, echoed the critique leveled against the mining west and toward all corners of the US mining industry since the late 1850s when Freiberg-trained mining engineers began returning with their newfound knowledge of European mining and milling practices. There was nothing overtly personal in these critiques; they represented a trend that would transform all areas of US life over the next several decades: the rise of the expert and the cult of the professional. While this transformation has been noted and studied in many areas of US history, it is often attributed strictly to social forces and is generally understood as having taken hold later in the 1870s, but the emergence of professional miners in the 1860s and their approach to the natural resources involved in mineral production suggest that there may be additional dimensions to understanding this trend. While Raymond expressed his professional critique as the cultural and social failure of western miners, at its foundation his view was rooted in a radically different mode of conceptualizing and understanding the natural world than the one the gold miners had. It was a new lens on nature that would ultimately have important consequences beyond Montana and the mining industry, although its expression there helps to highlight its main features.[41]

The rise of professional mining expertise involved a subtle shift in knowing nature, from an immediate knowledge derived from the obvious conditions that presented themselves to a removed or abstract knowledge by which the immediate conditions would be interpreted. Where the gold rush miners brought with them what might be called an agricultural or superficial perspective on the landscapes they encountered, the professional mining men thought in terms of what was underneath or behind, buried or concealed, what we might call a *subterranean lens*. For all of their chaos and uncertainty and the accidental outcomes, the gold rush miners confronted nature as it presented itself. Their relationship was physical and direct, and the knowledge they gained about mineral deposits derived from the work they did; as a result it was difficult, if not impossible, for them to transcend the limits of their own bodily senses in the experiences of their relationships to their claims and to the landscapes they sought to exploit. Professional miners, by contrast, worked in the abstract. They measured things to enlist knowledge from elsewhere to predict the future. Their relationship to mineral resources was based on reduction and distance, their knowledge derived from a dependence on the labor of others, but they appeared to provide a transcendent view of the mineral landscape and the challenges faced in the mining west.

These very different approaches to confronting and understanding the natural world produced, in the end, very different and even incompatible systems of mineral exploitation. The latter form had an advantage in that it produced a form of knowledge best suited for both the needs of a state interested in quantifying its resources and for the new forms of economic enterprise that sought similar levels of quantification in an effort to, as much as possible, eradicate uncertainty from the practice of mining. This way of knowing nature set down tentative roots in the 1860s and 1870s, but it would take more than the dogged efforts of Rossiter Raymond and the formation of a professional mining infrastructure for it to fully establish itself in the western mining industry.[42]

THE 1872 MINING LAW

As Montana miners and boosters licked their wounds and struggled to make sense of the lashing critique of their decade of work, back in Washington, DC, William Stewart—whose stopgap legislative effort in 1866 had also come under fire from Raymond and his cohort—began to set in motion revisions to the mining law that would have significant impacts on the future character of western mining, no less the trajectory of Montana's mining industry.

The 1866 mining law had been greeted with enthusiasm by western metal miners and their political supporters, but its enactment occurred just as professional miners were finding their voice. Unsurprisingly then, as a reflection of the culture of mining that had grown out of the gold rush period, Stewart's law quickly drew criticism. W. W. Keyes argued that the 1866 law was "a failure. The only good which, in my humble judgment, has been accomplished by the law of 1866, was the legalization of what would otherwise have been a trespass upon the public land of the nation." Writing from Montana Territory, where he had lived since 1864, Keyes had watched as Montana miners began to work ore lodes and establish districts out beyond the immediate hillsides of the placer mining creeks. The social conflicts that followed from these developments provided Keyes with the substance of his critique of the law.[43]

Based on what he saw in Montana, Keyes believed Stewart's law was "fatally deficient" for at least two reasons. First, it provided no clearly defined abandonment clause; men were making claims, accomplishing the minimum necessary work, and then going in search of other opportunities, Keyes complained. This had led to the tying up of potentially valuable natural resources in unworked but legally occupied claims. Unlike the placer gold district codes, which had required specific amounts of ongoing work to maintain title, lode claims—once located and provided with $1,000 in labor—could be (and were) held for an indefinite period. Where the lode miners, whose own interest was possession of locations, undoubtedly saw these actions as a form of investment, to the professional mining engineers they represented waste at least as serious as the poorly developed mines and wasteful gold-processing techniques of the placer miners.[44]

The 1866 mining law had also created an untenable policy with regard to surface rights. According to the law, the miner's patent was in a single stretch of a single ore lode, and it only allowed one lode per miner. This provision had been included to avoid the specter of monopoly, in which a mining company made claims on multiple sequential outcroppings in a mineralized zone. Its goal was to promote commercial interests in a just and democratic way, but, as Raymond quipped in 1870, "Nature has not regulated her operations by commercial rules."[45] By discouraging monopoly, the policy created situations where multiple claims to surface rights had been legally located on the same stretch of ground. Productive claims were as often as not found to contain more than one outcropping in close proximity, so, by rule, these other ore bodies could not be claimed by the initial claimant. They were thus available for other miners to patent, and those miners could make their own claims around them. Before long, lode districts began to contain multiple overlapping claims and unmanageable property rights conflicts. Rossiter Raymond dismissed the resulting policies guiding claimants as "one huge precipitate of amorphous opacity—voluminous instructions, learned contradictions and pompous dogmatisms, bobbing about in boiling confusion, and diluted, but not clarified, with no end of watery matter."[46] More succinctly and generously, mineral policy expert William Colby wrote, "Everyone recognized that the Act of 1866 had been hastily prepared and passed to meet an emergency and thus [to] forestall legislation hostile to the mining interest."[47] Time and again across the West, lode miners found themselves in conflict with other lode miners as a property rights policy rooted in the gold rush culture's knowledge of mineral resources could not quite reconcile itself with the variety of conditions present in ore lodes.[48]

By the early 1870s, the political climate in Washington had begun to change with regard to western minerals because of continued production of gold bullion since the passage of the 1866 law and a growing production of silver—another valuable coinage metal—at the same time. Stewart took the opportunity to completely rewrite the federal mining law and revise the way federal policy conceived of mineral resources. Stewart's new law, An Act to Promote the Development of the Mining Resources of the United States, was signed into law on May 10, 1872, by President Ulyssus S. Grant—superseding the 1866 legislation in its entirety.

The 1872 law, which remains in effect today, made every effort to avoid the sorts of conflicts that had emerged under the 1866 law, largely by altering its conception and understanding of mineral deposits to conform better to the knowledge of nature brought by professional mining engineers. Most obvious among the many changes was a redefinition of the nature of the resources to which the law applied, revising the surface-oriented term *mineral lands* in the 1866 law to the subterranean-oriented *valuable mineral deposits*.

The 1872 law made additional revisions to the procedures for locating claims and to the subsequent property rights. The 1866 law did not protect a miner's claim until he had marked the location, engaged in the required work, and filed the necessary paperwork with a district recorder. This made it possible for a discoverer who failed in any one of these steps to have his claim legally jumped, and such cases were not uncommon, especially in locations where valuable ore turned up. Under the 1872 legislation, in contrast, ore lodes and the claims that contained them became protected the moment an ore lode was discovered. Claimants would still have to file claims and the necessary paperwork, but they could be assured of legal protection if their claim was jumped before everything was completed.

The new law also increased the allowable length of claims along the mineral outcropping, which it referred to as the "apex," to 1,500 feet. This increase, like the change to a subterranean approach to mineral resources, reflected professional miners' notions of how much ore ought to be exploited by a single partnership. The 300-foot length defined in the 1866 law presumed that an individual miner or a small group of independent men would perform the work. The new allocation, five times longer, was a clear signal that companies and capital would be enlisted to exploit ore on these new claims and that more ore might be needed. The 1872 law also established a fixed claim width, requiring that 300 feet from the center of the apex on either side had to be included as the surface portion of the claim and that miners could purchase this surface ground outright for five dollars or less an acre, depending on the location. To avoid the conflict of multiple, overlapping claims, the 1872 law used the primary lode claim to dictate the surface rights and then allowed that the surface claim re-extended subterranean ore rights to the length of any sections of any other ore lodes whose outcropping, or "apex," appeared within the boundary lines of that claim. The one provision that did remain from the 1866 law was that miners had the right to follow their ore lodes into the ground as far as it went and in whatever direction it might dip, even if such a route put the miners' work underneath the surface claim of an adjoining property.[49]

In arguing before Congress for the passage of the 1872 law, Senator Stewart explained his reasoning for allocating property in subsurface ores that were potentially distinct from the surface claims. It is possible to see how the perspective of the European-trained miners and the subterranean imagination had found its way into his thinking and into federal law.

> Nature does not deposit precious metals in rectangular forms, descending into the earth between perpendicular lines, but rather in veins or lodes that vary from 1 foot to 300 feet in width, that dip from a perpendicular from 1 degree to 80 degrees, and that course through mountains and ravines at nearly every point of the compass. In exploring for vein mines, it is a vein or a lode that is discovered, not a quarter-section of land marked by surveyed boundaries. In working a vein, more or less land is required, depending on its size, course, dip, and a great variety of other circumstances impossible to provide for by passing general laws. Sometimes these veins are found in groups within a few feet of each other and dipping into the earth at an angle of 30 to 50 degrees, as at Freiberg, Saxony, and Austin, Nevada. In such cases a person who buys a single acre in a rectangular form would have several mines at the surface and none at 500 or 1,000 feet in depth. With such a division of a mine—one person owning it at the surface and another at a greater depth—neither owner would be justified in expending money on costly machinery, deep shafts, and long tunnels for working the mine.[50]

Despite the fact that very few western mines had even approached 500 feet in depth and none had burrowed as deep as 1,000 feet, Stewart was convinced that his law would "prevent litigation and give certainty to mining enterprises."[51] No doubt feeling justly responsible for the significant changes made to the mining law, a few years later Rossiter Raymond looked back on its passage and celebrated the changes it had wrought: "Its immediate effect was to encourage the adventurous and speculative exploration of one new district after another; and, as the pioneer industry of mining blazed the way for the advance of all other civilized activities (though too often at a ruinous cost to itself), the result was the conquest of a vast wilderness, and the creation of a new empire."[52]

MONTANA'S SILVER RUSH

In August 1872, one year after Raymond and Eilers visited Montana and three months after the passage of the 1872 mining law, William Andrews Clark rode a stagecoach 500 miles south from Helena to Ogden, Utah, where he boarded a train for a five-day rail trip to New York City. Among his belongings were several sacks of ore—samples from each of four silver mines he had recently purchased in the Summit Valley and relocated under the requirements of the new mining law. Clark was on his way to the School of Mines at Columbia College to enroll in a semester of courses in mine engineering. The federal government had not followed Raymond's 1869 suggestion to the US Congress that it establish a national school of mines, but private colleges recognized the opportunity that presented itself after the Civil War. By 1872 Columbia College in New York City had cobbled together a faculty and launched one of America's first mining schools, which, in addition to a full two-year program of study, also offered courses by the semester for students who might not need the full degree.[53]

Clark had been in the northern Rocky Mountains since the second year of the gold rush in 1863, when he and a partner had joined a large wagon train of men migrating from the Colorado gold mines. Clark had tried a little mining but soon turned to the more lucrative businesses of freighting supplies, carrying mail, and marketing gold. By the early 1870s he had amassed a sizable fortune—owning a bank and a wholesale merchandizing and mining supply store in Deer Lodge; investments in real estate in Butte, Missoula, and Helena; and several silver mines and gold quartz claims around the region. In 1872, in the wake of the new mining law and perhaps in reaction to the negative report by Raymond and Eilers, Clark set his attention and capitalist energies on the ore lodes in the Summit Valley, above the abandoned Silver Bow Creek diggings adjacent to the tiny settlement named Butte.[54] There, a series of parallel outcroppings, some wider than fifteen feet across, arched across the surface of the hill from southeast to northwest. Mines close to the ones he had located had produced an ore that would not reduce, even in the very hot temperatures of the homemade furnace constructed by William Parks in 1865 shortly after the Silver Bow District was first founded by placer miners. Clark believed these ores must be worth something; if so, he hoped to learn what kind of processing was necessary to render their value.[55]

As Raymond and the professional mining engineers had argued, exploiting more complex mineral compounds required new approaches and new technologies, and it appeared that this message got through to Clark. Clark's behavior in 1872 and again in 1875, following the 1873 economic depression, displayed a decidedly systematic and conservative approach to mineral development. Clark learned at the School of Mines that some of his ores were indeed rich in silver—rich enough, in fact, to pay for the extraction of the highest-grade ores and their shipment by stagecoach to the railheads in Corinne, Utah, and from there east for processing, at a profit. With this information, Clark returned to Butte in the spring of 1873 and began developing his mines, sending the richest ore to Baltimore, where a new smelter based on Swansea designs extracted its silver. Clark also began to build a stockpile of his lower-grade ore, which could also be shipped east at a profit as soon as the railroad and its lower transportation costs arrived.[56] Construction was under way on the Northern Pacific Railroad being built west from Duluth, Minnesota—following the path once trod by the Fisk Trains—across the high plains of the Dakotas, aiming for Montana and the northern Rocky Mountain mineral region. Construction had also begun on a Union Pacific spur north from Corinne, Utah, following the route of the old Corinne Road through southern Idaho and into Montana Territory. But the national economy had other plans. A general depression struck in the fall of 1873 as railroad stocks collapsed in the wake of Jay Cooke's announced bankruptcy. Railroad building and silver mining, among dozens of American industries, ground to a halt.[57]

When the economic depression began to lift in late 1874, efforts to develop successful lode mining on the Butte Hill ore lodes began anew. After trying various means to work the ores he had taken from his Asteroid Mine on Butte Hill during the late 1860s and early 1870s, William Farlin abandoned his claim in disgust in 1874 and set out for Idaho. Unable to give up completely, though, he had brought along samples of the ores. Late that year in Owyhee, Idaho, Farlin learned from a skilled assayer that his Butte ores were rich in silver and could be worked profitably in a modified stamp mill. Farlin returned immediately to Butte, borrowed $30,000 from William Clark, and in 1875 began construction on a mill he named the Dexter Mill.[58]

Farlin's discovery that he held valuable ores, and his subsequent launching of the Dexter Mill construction, encouraged other claim holders on Butte Hill, several of whom excavated mine shafts and removed ore samples to see if they could stimulate an interest in their claims among mine developers elsewhere in the West. Farlin was

William Andrews Clark, 1898. Eventually rising to national significance as, first, a seated and then an unseated US senator and copper king, William Clark came to Montana during the 1863 gold rush and initially made his money transporting gold and mail in and out of Montana Territory's placer mining centers. Photograph by Wilhelm, New York City; courtesy, Montana Historical Society Photography Archives, catalog #941-722, Helena.

unable to complete the mill with the borrowed money, and he failed to pay his debt; as a result, Clark took over the operation in early 1876. That same year, after examining some of the Butte Hill ores, the Walker brothers of Salt Lake City, Utah—western mining capitalists who had successfully developed hard-rock gold and silver mines in Nevada, California, and Utah—sent a trusted superintendent, Marcus Daly, to Butte to examine the silver mines there. Daly's experience in the silver mines of the Comstock Lode and in Utah helped him recognize that these ores were profitably rich in silver, and his familiarity with a variety of ore-processing techniques helped him understand what kind of mill he would need to treat them. On Daly's advice, the Walkers purchased two large claims along an ore body, eventually named the Rainbow Lode for its colorful ore. Daly remained in Butte to superintend development of these claims, named the Alice group by the Walkers, and to oversee the reconstruction of a stamp mill the Walkers had sent north from Utah.[59]

Also in 1876, Andrew J. Davis—who, like Clark, had made a small fortune providing merchandise for the gold miners and who was clearly spurred into action by Davis's and the Walker brothers' investments—began developing his Lexington Mine on Butte Hill and constructing a processing mill of a design similar to theirs. By the beginning of 1877, Butte Hill had suddenly emerged as a silver production center, with several mines in development and hillsides populated with processing mills connected to increasingly well-developed ore lodes.[60]

None of the new mine developers in Butte's silver industry possessed the full range of mining training Rossiter Raymond had hoped Congress would work to cultivate in the US West, but each had taken the steps the mining engineers had recommended, developing their mills only after determining the kind of ore their mines contained and developing or hiring skills where theirs were lacking. Among other things, these men had learned that the apparently complex ores in these mineral deposits were a silver chloride. The ore looked a lot like a smelting ore, but it was made up of a compound that contained its silver mechanically rather than chemically. These were what were known as *free-milling silver ores*, whose development and production required essentially the same treatment as that of gold quartz: stamping and amalgamation, nothing else. The first developers on the Butte Hill silver lode got a leg up by being able to see beyond the surface appearances of their ore and project their valuable holdings into the earth.

Yet for all its differences from placer gold mining, in the end lode mining had the same ironic material conditions in which it persistently worked against its own reserves. Mining men and partnerships organized capital to coordinate the extraction and processing of valuable ores, but the more ore that was removed and stamped, the less ore remained in the ground to be mined. For placer gold miners, this fact led them to work over the region of their claim and then move on. For lode miners, it meant they had to push their mines deeper and deeper into the earth to access more of the ore lode, if it existed at all. This inexorable push downward, as lode miners across the West had begun to learn, was not merely a matter of digging deeper; it also meant encountering new kinds of conditions underground. Indeed, as it turned out in Butte, at depths of anywhere between 60 and 180 feet, mines encountered groundwater. "A dry mine may have persistence of ore to indefinite depth and its engineers can plan accordingly," Otis Young wrote, "but a wet mine is in trouble from the first day water is encountered."[61]

The well-funded Alice Company, led by the ambitious Marcus Daly, was the first of the new silver companies to encountered groundwater in its mines. "Having encountered another strong flow of water in the main shaft of the Alice Mine, Rainbow lode, last week," the *Engineering and Mining Journal* reported in 1877, "Mr. Daly was compelled to suspend the work of sinking, the water was coming in so rapidly as to make it impossible to go down the shaft without other hoisting works." Daly's miners had reached 140 feet into the ground. A few days later "another vein of water [was] struck in the cross-cut about two feet from the shaft, and the water came in so rapidly that all work at the depth had to be abandoned. It will be impossible to do anything below water level without a pump," the *Butte Miner* reported. The Lexington Mine was not far behind. That same year, 110 feet below the surface, miners working the Lexington Mine struck "a good flow of water, so that pumping or hoisting machines of considerable power will be needed."[62]

For the Alice Company, the most highly capitalized of the new silver companies, this difficulty was overcome by additional investment in the mine. Daly ordered two enormous Knowles pumping engines—one for immediate use and one for future development—which arrived a few months later. Once installed, the pumps began pulling the water out of the mine faster than it could pour in, allowing Daly's miners to return to work at the new depths and continue to push downward. No working mine in the district would be able to avoid this ubiquitous groundwater.[63]

But a single water pump of this kind cost upward of $50,000 installed in the late 1870s, not including the cost of fuel, replacement parts, general maintenance, and loss of equipment if the machine broke down and the mine re-flooded. In hard-rock mining, much more than in placer gold mining, hard labor and common sense could not overcome all of the obstacles to knowledge. A mine full of water could not be emptied with buckets. The only choice was to use the brute force of high-powered pumps: solutions to groundwater required machines that cost money. Daly had recycled a mill for the Walkers and developed enough of the upper parts of the Rainbow Lode to justify paying for the investment in an expensive pump, and the Alice Mine resumed work within months of striking groundwater. The same was true for the Lexington Mine, where A. J. Davis—the primary importer of mining machinery for Butte in the 1870s—installed whatever equipment he needed. In contrast, mines such as the Stevens, known locally as "the best poor man's mine in Butte" and owned and operated by a group of local miners, found the intrusion of underground water a problem beyond their reach. As the *Engineering and Mining Journal* reported on the Stevens in early 1881, "The work of developing has been attended with considerable difficulty; the water coming in necessarily made the work go slow." The inability to purchase adequate pumping machinery literally sunk the hopes of the Stevens's "owners." While the Alice and the Lexington were developed into valuable silver mines in the 1870s, the Stevens would have to sell out to Marcus Daly before it produced valuable ore.[64]

Encountering groundwater not only significantly raised the costs of dead work, it also presented new risks for the mining companies. Because of water's solvency, the ore lodes could become valueless, or they could carry a different mix of mineral compounds in the saturated ground. Reaching groundwater posed a fairly significant gamble for hard-rock developers. They could not know the condition of the ore lode without gaining access to it, but access was expensive. With their pumps, Daly and Davis had merely created conditions whereby the work of opening vertical shafts and digging a horizontal tunnel to the face of the ore could continue. But this guaranteed nothing, and significant investment in western mines had been lost when similar gambles did not pay off.[65]

Butte's silver mines were a good gamble, as the ores changed from silver chloride to a silver sulfide in the saturation zone, but this could not be known without direct contact with the ores themselves. The Alice Company's early purchase of a Knowles pump also made it the first to assess these ores. In March 1878, after groundwater removal and several weeks of expensive dead work tunneling to the face of the ore lode underground, the miners finally brought samples of the new ore to the surface where it could be assayed. The *Butte Miner* reported the promising results almost the moment they were in. "At the depth of 16 ft. below the lowest level of the mine, the shaft entered a chute of ore, giving an average assay of 40 ounces of silver to the ton," the *Miner* wrote. "This ore is base, but can be worked by roasting, and is quite rich enough to pay all expenses and a fair profit besides." Within a year of the Alice Company's assessment of the saturated ore, other mining companies on Butte Hill with the means to invest in pumps did so at a far diminished risk.[66]

Having the means to pump out water from the mines and finding silver-enriched ore in the saturation zone had the effect of proving the value underground at the same time it diminished the value of the capital investments on the surface. Specifically, the Dexter and Lexington stamp mills could no longer do the work of reducing ores. The sulfide ore in the saturation zone held the silver in a chemical bond, so it could not be mechanically stamped out of the ore unless the ore was first converted by roasting. This meant that the underground success led to the need for more investment on the surface, in the form of the construction of a furnace in which the ores would be mixed with an equal part of salt to facilitate the burning off of the sulfur and convert the remaining material to a silver chloride, which could then be stamped and amalgamated. In this way, hard-rock developers found themselves caught in a new pattern of mineral exploitation whereby expensive investments in their mines led to a need for expensive new processing technologies on the surface and a new calculus of scale to accommodate the increased capital investment.

The expense of adding a roaster seemed to have encouraged the developers to increase their overall milling capacity concurrent with roaster installation. The higher cost of keeping a mine open below the water table gave developers a strong incentive to work more ores in an effort to achieve an economy of scale and recover their added expenses more quickly. Ironically, however, the developers built new mill capacity that exceeded their rate of ore production. One developer, for example, found that it could only run at half capacity after it opened its new mill, and the Alice Company found it necessary to hire out prospecting teams to the independent mines on Butte Hill to ensure an adequate quantity and quality of ore for its new roasting and milling works.[67]

All of this mining and milling and processing, for which the region showed a steady increase after 1876, intensified demands for other regional resources, especially timber. The various shafts that led from the surface into the mines were usually completely closed in with timber, and the tunnels leading to the ore faces were lined with timber as well. In the drifts of the ore lodes, Philip Deidenheimer's square-set timbering technique was most commonly used to hold open the space left empty by the excavated ore lode. After examining the new shafts of the Gagnon and Original Mines, which had reached groundwater at the unusually shallow level of fifty-seven feet in 1878, a *Butte Miner* reporter noted that "the different drifts are all timbered in the most substantial manner." The new mills and roasting buildings were also consumers of lumber, as they were constructed entirely of wood. The demand for timber soon exceeded its supply in the region. In 1881, for example, the owners of the Alta-Montana's mill found themselves delayed a month in finishing its reconstruction because of "a scarcity of timber." A similar situation had confronted the Alice Company during the summer of 1880. The lumber production operations constructed to serve placer mining during the 1860s could not produce the quantity of timber necessary for the growing scale of hard-rock development in Butte by the 1880s, and no new lumber industry immediately filled the void.[68]

The changes in the region's landscape, the new mills and roasters—especially in Butte—and the increasing number of shaft houses and engine rooms along the ore lodes of Butte Hill were more than cosmetic changes. They were also social spaces, and these new forms of production also heralded a reshaping of social relationships in the Summit Valley mining region—a location quickly emerging as one of the region's most significant silver production centers. In the spring of 1878, the *Butte Miner* reported that the independent owners of the Mountain Boy Mine were "dissatisfied with the prices demanded for working its ores at the Butte mills." These men represented a mining strategy that was being priced out of the market in 1878, an under-capitalized partnership in which the mine owners also worked their claim. For a few years, these kinds of partnerships made claims and supported their work by selling their ores to one of the private mills or to the custom mill and smelter built in Glendale in 1877, but the companies that added roasting and expanded processing capacity sought every means to cut costs—including paying as little as possible for the labor of others. The Mountain Boy miners had reached fifty feet into the ground that year and hoped to finance the doubling of that depth during the upcoming summer through the sale of their ores. The silver sulfide ore they pulled from their mine yielded an average of 100 ounces of silver per ton of ore removed. In New York City, that material fetched $1.18 an ounce. Mill owners in Butte, however, accounting for their various costs, set the Butte price at a fraction of a cent under 50 cents an ounce—less than half the New York rate. That price did not return enough to the Mountain Boy miners to fulfill their ambitions, but their need for cash forced them to grudgingly accept the terms. Later that year, when Marcus Daly announced the Alice Company's intention to lower wages in the company mines by 50 cents a day, the announcement sparked a walkout that gave birth to the first hard-rock mining union in Montana: the Butte Miners' Union.[69]

THE LEGACY OF LODE MINING

Within three years of its full-scale launch in the Summit Valley, lode mining designed on the principles and practices suggested by professional mining engineers had developed an unexpected dynamic: the continued exploitation of ore did not lead to stability and permanence but instead seemed to generate a pattern of iterative reactionary growth and social change. The ubiquitous presence of groundwater required a major expense in water pumps and exposed new kinds of ore whose processing required new capital investments in the mills and whose expansion encouraged the enlargement of processing capacity. The new mill sizes, in turn, encouraged an expansion of work underground, a higher demand for resources, and an effort to minimize labor costs. Changes in one factor led to a ripple of changes across the face of the lode mining system.

One might be tempted to conclude that this dynamic was caused by the engagement of new strategies and techniques brought to the West by the Freiberg men and legally facilitated by the revised 1872 mining law, but that would oversimplify a much more complex set of systemic relationships. To begin with, as with placer mining, the radical uncertainties of ore deposition could not be eliminated; in contrast, lode mining did allow mining companies to evaluate, and to some degree estimate, the value and extent of their ore in advance of a full commitment to a particular mining claim. The subtle marginal difference in predictability at once elevated the sense of control—which was particularly nonexistent in placer mining—and also elevated the overall cost of mining itself, as knowledge and technology were not free and could not be attained through simple experience. Together, these factors contributed to the confidence to make investments and intensify the practice of lode mining. These investments and intensifications changed the nature of lode mining such that while the original silver lode operations in Butte may have appeared somewhat similar to the first lode mining done in the Dakota claim at Bannack, they in fact represented a new, emergent form of the practice whose dimensions did not become clear until the 1880s.

As these new lode mining operations began to draw upon the knowledge and techniques developed in the European centers, they deployed this intelligence and practical knowledge within a context fundamentally different from the one in which they had been created and enlisted that intelligence and knowledge within a social structure substantially different from the one in which they had originated. The differences are important, as they help explain the sudden acceleration of growth in scale that marked the 1880s. The knowledge and technique transferred to the mining west after the 1860s essentially constituted a cultural and, in the case of Germany, public entity. When that entity was resettled in the US West, it came as a commodity, an essentially economic entity that could be bought and sold on the market as needed. The unmooring of technique from culture began with Germany opening its Bergakademie to paying American students—which, if the experience of many of the early students is any indication, was not done enthusiastically—but it found its expression in the US metal mining industry after the 1860s. Even Rossiter Raymond seemed to recognize the cultural quality of the knowledge he had attained, and he made a sincere and concerted effort to institutionalize its domestic production under the state in the mining west. Instead, mining knowledge was destined—like more and more aspects of life in the United States in the years ahead—to be traded on the market.

The second difference, which is distinct from but related to the first, is found in the kinds of social structures or institutions that controlled and put this knowledge to work. In Germany and Wales, for all their differences, the mines and mills were organized to produce metal—lead and silver in Germany and copper in Swansea. The social goals, in other words, were to contribute a "use value" to the national and world economy. The slow development and systematic design in Saxony and the secretive and almost cultish organization around Welsh smelters could both be attributed to this structure. In the US West, by contrast, the goals of the institutions that adopted and enlisted the new knowledge had changed very little from those of the original placer gold seekers: the production of metal was secondary to the aim of making money. These were institutions of capital accumulation; they produced metal for its "exchange value."

Together, these factors contributed to the dynamics of the lode mining system that began to settle into place in the US West in the 1870s. The path had been blazed by the work of placer gold miners, but the combination of imported production strategies and knowledge, generous federal policy-making, organized private partnerships, and the existence of generally high-value ore lodes in locations like Butte Hill crystallized a new form of mining in the region by the late 1870s. Among its most indelible characteristics was the propensity to expand, generated out of the complex interactions among the conditions of the ore as miners pushed deeper into the ground; the costs of excavation, new equipment, and labor; and the organization of the entire undertaking around the goal of profit. For Butte Hill, this dynamic led to phenomenal growth in both the scale and number of mining enterprises, as it became increasingly apparent that the collection of large outcroppings was unusually rich and, at least by the early 1880s, had yet to show significant decline in ore quality with depth.

By the end of the 1870s, a large quantity of valuable ore had been removed from the earth beneath Butte Hill, but it was no longer just silver being profitably pulled from the ground. In the central ore zone of Butte Hill, where surface ores held silver, gold, and copper, crossing the water line had brought even more surprising changes. "Some of the ores taken from below the water-line may be milled after being roasted," Z. L. White wrote as these ores were uncovered, "but those containing galena in considerable quantities, manganese, and copper can only be treated by fire." Precipitation of water from the surface through these ores had created a zone of ore with pockets of copper unprecedented in its richness, a copper "saturation" zone but with a single, crucial, catch: the ores were predominantly copper-iron sulfides that could only be reduced through the intense oxidation processes of smelting.[70]

NOTES

1. Gleb S. Pokrovski and Leonid S. Dubrovinsky, "The S³ Ion Is Stable in Geological Fluids at Elevated Temperatures and Pressures," *Science* 331, no. 6020 (February 25, 2011): 1052–54; Chil-Sup So and Kevin L. Shelton, "Stable Isotope and Fluid Inclusion Studies of Gold- and Silver-Bearing Hydrothermal Vein Deposits, Cheonan-Cheongyang-Nonsan Mining District, Republic of Korea, Cheonan Area," *Economic Geology and the Bulletin of the Society of Economic Geologists* 82, no. 4 (July 1, 1987): 987–1000.

2. See Paul, *Mining Frontiers,* esp. chapter 4, "The Comstock Lode, 1859–1880," 56–86; also Rickard, *A History,* esp. chapter 5, "The Comstock Lode," 82–114.

3. Browne, *Report,* 498; Fisk, *Expedition,* 28; Eaton et al., "Notes on Montana," 141; Raymond, *Silver and Gold,* chapter 4, "Montana." This chapter includes a complete survey by W. S. Keyes and others of Montana's mining prospects.

4. Browne, *Report,* 498; Fisk, *Expedition,* 28; Eaton et al., "Notes on Montana," 141; Raymond, *Silver and Gold,* chapter 4, "Montana."

5. See Curtis H. Lindley, *A Treatise on the American Law Relating to Mines and Mineral Lands within the Public Land States and Territories and Governing the Acquisition and Enjoyment of Mining Rights in Lands of the Public Domain,* 3rd ed., 3 vols. (San Francisco: Bancroft-Whitney, 1897), esp. 57–70. See also Swenson, "Legal Aspects," which suggests the three-policy description. James E. Wright, *The Galena Lead District: Federal Policy and Practice, 1824–1847* (Madison: Department of History, University of Wisconsin, 1966), makes the intriguing argument that the western mining districts were based on the rules of federal policy written for Illinois in the early nineteenth century.

6. Lindley, *Treatise,* 61–71.

7. Ibid.

8. 39th Congress, Sess. 1, chapter 262, July 26, 1866; Colby, "Origin and Development," 463.

9. Quoted in Colby, "Origin and Development," 458.

10. E. F. Dunne, "The United States Mining Law," *Engineering and Mining Journal* 9 (1870): 2.

11. These statistics were compiled from data assembled in sequential Treasury reports. See Eaton et al., "Notes on Montana"; Rossiter W. Raymond, *Statistics of Mines and Mining in*

the States and Territories West of the Rocky Mountains, 1869 (Washington, DC: Government Printing Office, 1870); Rossiter W. Raymond, *Statistics of Mines and Mining in the States and Territories West of the Rocky Mountains, for the Year 1870* (Washington, DC: Government Printing Office, 1872); Rossiter W. Raymond, *Statistics of Mines and Mining in the States and Territories West of the Rocky Mountains, for the Year 1871* (Washington, DC: Government Printing Office, 1873).

12. Malone, *Montana*, 141–42; Phillips, *Forty Years*, 33.

13. Spence, *Mining Engineers*; Raymond, *Mineral Resources . . . 1869*, 236–37.

14. William Henry Pettee of Massachusetts attended the Bergakademie immediately following the Civil War and documented his experiences in detailed letters home. See the Letters of William Henry Pettee to Matilda (Sherman) Pettee, 1867, Folder 3, Box 2, Pettee (William H.) Collection, HHMC; John R. Leifchild, *Cornwall: Its Mines and Miners with Sketches of Scenery Designed as a Popular Introduction to Metallic Mines* (London: Longman, Brown, Green, and Longmans, 1855).

15. William Henry Pettee letter. See also J. C. Bartlett, "American Students of Mining in Germany," *Engineering and Mining Journal* 23 (1877): 257–58.

16. Benjamin Lyman, "The Freiberg School of Mines," chapter 34 in Raymond, *Mineral Resources . . . 1869*, 236–38.

17. "Report on the Mineral Property of the Germania Gold Company in Griffith District, Clear Creek, Co., Colorado Territory," 5, June 20, 1864, Folder "Colorado," Box 129E3, New York Public Library [hereafter NYPL]; "Report on the Manhan Mine, South Hampton, Mass.," 2, November 14, 1865, Folder "Massachusetts," Box 129E3, NYPL.

18. Rossiter W. Raymond, *American Bureau of Mines, Prospectus* (New York: Wm. C. Bryant, 1866), 5, Folder "Miscellaneous Items," Box 129E3, NYPL. See *American Mining Journal* 7 (1869) and *Engineering and Mining Journal* 8 (1869), especially late June and early July, for editorial discussions of these changes.

19. Rossiter W. Raymond, *Biographical Note of James Duncan Hague: A Paper Read before the American Institute of Mining Engineers, at the Chattanooga Meeting, October, 1908* (Chattanooga: Author's ed., 1909); "Letter of Instruction from Hugh McCullough," in Raymond, *Mineral Resources*, 1.

20. James D. Hague, with geological contributions by Clarence King, *Mining Industry*, vol. 3 of the United States Geological Exploration of the Fortieth Parallel (Washington, DC: Government Printing Office, 1870). See also Raymond, *Biographical Note of James Duncan Hague*, 2–4.

21. See "Letter of Instruction, August 2, 1869," in J. Ross Browne and James W. Taylor, *Reports upon the Mineral Resources of the United States* (Washington, DC: Government Printing Office, 1869), 4–5; Raymond, *Mineral Resources 1869*.

22. Raymond, *Mineral Resources 1869,* 5.

23. Raymond, *Mineral Report for 1868*, 5–6.

24. Ibid.

25. Ibid.

26. Raymond, *Mineral Resources*, 175.

27. Raymond's comment was made in reaction to Rothwell's paper, and he typed and amended his comment to the essay in the volume of the transactions. Richard P. Rothwell, M.E., "Remarks on the Waste in Coal Mining," in Rossiter W. Raymond, ed., *Transactions of the American Institute of Mining Engineers*, vol. 1, May 1871–February 1873 (Philadelphia: AIME, 1873), 56; Rossiter W. Raymond, "The National School of Mines," *Engineering and Mining Journal* 9 (1870): 184. Raymond complained frequently about the decreasing funding and inadequate distribution of his survey. See, for example, Rossiter W. Raymond, "Mining Statistics," *Engineering and Mining Journal* 9 (1870): 185.

28. Raymond, *Mineral Resources*, 175.

29. Ibid., 177, 216, 229.

30. The *Engineering and Mining Journal* "is the organ of the American Institute of Mining Engineers and is regularly received and read by all the members and associates of that large and powerful society, the only one of its kind in this country. It is therefore the best medium for advertising all kinds of machinery, tools, and materials used by engineers and their employees." First entry in the *Engineering and Mining Journal* 11 (1871): 5. See Raymond, *Transactions of the American Institute of Mining Engineers*, vol. 1, May 1871–February 1873, and subsequent volumes.

31. "To Whom It May Concern," *New North-West* [Deer Lodge, MT], September 17, 1869, 1; Rossiter W. Raymond, "To Whom It May Concern," *Engineering and Mining Journal* 8 (1869): 135.

32. Raymond, "To Whom It May Concern," 135.

33. Rossiter W. Raymond, "Editorial Correspondence: The Dormant Resources of Montana; Virginia City, Montana Territory, July 29, 1871," *Engineering and Mining Journal* 12 (1871): 13.

34. Rossiter W. Raymond, "Editorial Correspondence: Montana and the Railroad; August 2, 1871," *Engineering and Mining Journal* 12, no. 8 (1871): 153–54: "The merchants were frequently obliged to give credit to miners, who are wintering, without work, or stripping and preparing their claims. When they begin to clean up from their sluices, if they are successful, he [the merchant] gets his money, if not, he must write the account to profit and loss, or wait for his debtors to 'strike it rich' another season."

35. Rossiter W. Raymond, *The Mines of the West: A Report to the Secretary of the Treasury* (New York: J. B. Ford, 1869), 150.

36. Anton Eilers, "Western Montana, Deer Lodge City, August 6, 1871," *Engineering and Mining Journal* 12 (1871): 168, 191.

37. Anton Eilers, "Silver Smelting in Montana, I: General Considerations," *Engineering and Mining Journal* 12 (1871): 241.

38. Raymond, "Editorial Correspondence: Montana and the Railroad," 154.

39. Ibid., 281.

40. Raymond, *Silver and Gold*, 281.

41. Spence, *Mining Engineers.*

42. James C. Scott, *Seeing Like a State: How Certain Schemes to Improve the Human Condition Have Failed* (New Haven, CT: Yale University Press, 1998), esp. Part 1, "State Projects of Legibility and Simplification."

43. William W. Keyes, "The Defects of the Mining Law," in Rossiter W. Raymond, ed., *Statistics of Mines and Mining in the States and Territories West of the Rocky Mountains, 1873* (Washington, DC: Government Printing Office, 1874), 513.

44. Ibid.

45. "Mineral Deposits," *Engineering and Mining Journal* 10 (1870): 447.

46. Rossiter W. Raymond, "The New Mining Law," *Engineering and Mining Journal* 11 (1871): 121.

47. William E. Colby, "The Extralateral Right: Shall It Be Abolished?" *California Law Review* 5, no. 1 (1916): 18.

48. Lindley, *Treatise*, 74–80, 94–104; Section 3 of the Mining Law of 1872, quoted in "The Mining Law," *Engineering and Mining Journal* 11 (1871): 122.

49. The 1872 law fixed some of the problems identified in the 1866 law, but it also created new ones. See Rossiter W. Raymond, "Preface," in *Transactions of the American Institute of Mining Engineers*, vol. 1, May 1871–February 1873, iv.

50. Quoted in Colby, "Origin and Development," 456.

51. Quoted in Colby, "Federal Mining Act of 1872," 20.

52. Rossiter W. Raymond, "Comparison of Mining Conditions To-day with Those of 1872, in Their Relation to Federal Mineral Lands," in *Transactions of the American Institute of Mining Engineers*, vol. 48, February 1914 (New York: AIME, 1915), 303.

53. Glasscock, *War of the Copper Kings,* 62; Malone, *Battle for Butte*, 15.

54. Malone, *Battle for Butte*, 13–15; William S. Greever, *The Bonanza West: The Story of the Western Mining Rushes, 1848–1900* (Norman: University of Oklahoma Press, 1963), 223. See also Michael P. Malone, "Midas of the West: The Incredible Career of William Andrews Clark," *Montana, the Magazine of Western History* 33, no. 4 (1983): 2–17.

55. William Parks described his experiences trying to smelt copper in Butte in *Transcript of Testimony,* vol. 1, 194, Colusa-Parrot Mining & Smelting Company v. Anaconda Copper Mining Company, #61, Civ. Case Files 1890–1912, US District Court, Butte, Record Group 21, National Archives Pacific Alaska Region, Seattle, WA [hereafter NARA-PR].

56. Robert George Raymer, "A History of Copper Mining in Montana," PhD diss., Northwestern University, Evanston, IL, 1930, 9.

57. Malone and Roeder, *Montana*, 130–31, 140–41.

58. Otis E. Young, *Western Mining: An Informal Account of Precious-Metals Prospecting, Placering, Lode Mining, and Milling on the American Frontier from Spanish Times to 1893* (Norman: University of Oklahoma Press, 1970), 199; Malone, *Battle for Butte*, 15; Rickard, *A History*, 348; Greever, *Bonanza West*, 222.

59. Malone, *Battle for Butte*, 20; Rickard, *A History*, 349.

60. Raymer, "History of Copper Mining in Montana," 12; Young, *Western Mining*, 198.

61. Young, *Western Mining*, 80.

62. Excerpts from the *Butte Miner*, included in "Mining Notes: Montana," *Engineering and Mining Journal* 24 (1877): 379, and 23 (1877): 320.

63. "Mining Notes: Montana," *Engineering and Mining Journal* 30 (1880): 10, 302, and 31 (1881): 78.

64. Young, *Western Mining*, 171: "It is well to note that the claim that Cornish pumps were uncommonly economical to operate (nine cents per ton of water raised) was usually founded only upon analysis of fuel consumption while disingenuously omitting such matters as capitalization and installation, maintenance, and general losses to the operation occasioned by down time for frequent overhaul and major repair." "Mining Notes: Montana," *Engineering and Mining Journal* 31 (1881): 200, 217 (quote in the text).

65. Edward Dyer Peters Jr., "The Mines and Reduction Works of Butte City, Montana," in Albert Williams Jr., ed., *Mineral Resources of the United States, Calendar Years 1883 and 1884*, Report for the Department of the Interior (Washington, DC: Government Printing Office, 1885), 381; Young, *Western Mining*, 150.

66. Excerpts from the *Butte Miner*, included in "Mining Notes: Montana," *Engineering and Mining Journal* 25 (1878): 205; Tuberose, "Montana Mining Notes," *Engineering and Mining Journal* 27 (1879): 336.

67. "Mining Notes: Montana," *Engineering and Mining Journal* 24 (1877): 327. "The small [Glendale] furnace has yet been used. The large one has been finished and ready for work for some time, but the 50 to 60 tons of ore necessary to keep it running could not be delivered," "Mining Notes: Montana," *Engineering and Mining Journal* 27 (1879): 336.

68. "Mining Notes: Montana," *Engineering and Mining Journal* 25 (1878): 205, and 28 (1881): 330.

69. Emmons, *Butte Irish*, 226; "Mining Notes: Montana," *Engineering and Mining Journal* 25 (1878): 205.

70. Z. L. White, "The Mines around Butte City," *New York Tribune*, October 20, 1879.

Our comparatively undeveloped deposits, the increasing demand, our cheap food, the cheapness of the voyages across the Atlantic, the low cost of land, the intelligence of our workmen, and our labor-saving machinery ought to enable our industrial [mining] establishments, if properly located and conducted in the best manner—that is, with a view of obtaining the greatest product, of the best average, quality, and at the lowest possible price— to compete successfully, before many years have rolled around, for the markets of the world.

Eckley B. Coxe, "Mining Engineering" (1878)

When after a time the peaceful conquest of the forces of Nature in the New World halted and languished for lack of a sufficient supply of the sinews of war, the discovery of alluvial gold in California and Australia electrified two continents, revived the drooping industries of the World, whitened the seas with sails of commerce, brought the nations nearer together and forged another link in the chain of universal brotherhood which shall eventually girdle the earth. And later the discovery of rich veins of gold and silver bearing rock in place, alone, made it possible for the American people to recover from the era of inflation which the necessities of war produced, and to resume their onward march without an instant's confusion or delay on the solid terra firma of specie payments. And now when Science reveals the presence all around and about us, of a mighty force in Nature, glimpses of whose terrible power we have often witnessed with awe, in the lightning, the tempest and the electrical storm, at whose dread approach all Nature stands hushed and in whose track, death and desolation reigns supreme; when Science discovered that this mighty power, so destructive when unbridled, could under certain conditions, and within certain limitations be controlled, directed, and made to do the bidding of man, almost simultaneously with this discovery, vast and apparently inexhaustible supplies of copper were uncovered where they had lain buried for years, in the heart of the Rocky

THREE

Turning Copper into Gold

The Dynamics of System Building

DOI: 10.5876/9781607322351.c03

Mountains, which supplies, added to those already produced elsewhere, are barely sufficient when combined to meet the world's constantly increasing demand for this metal with which to harness the new steed, Electricity, and make it subservient to the uses of man.

F. E. Sargeant, *Report on Anaconda* (1895)[1]

THE DEVELOPMENT OF COPPER RESOURCES in the mountain west heralded a new age of mining, not merely for the western mining region but for the nation and even the world. By the late 1870s the fleeting days of gold were long gone, and the tumultuous days of silver were passing their apogee. During the 1880s, copper production increased to more than fill the void.

Before the 1880s, few people dreamed about mining copper in the West, never mind trying to produce the metal for a profit. During the entire nineteenth century, world copper production had been dominated by the Welsh smelters in Great Britain, although "native" copper deposits in the Keweenaw Peninsula, Upper Michigan—which connected to the water route east through the Great Lakes and the Erie Canal—gave Michigan producers a growing slice of the copper market after the Civil War. But Michigan copper had the advantage of requiring only mechanical processing, needing little more than specialized stamp mills to separate the pure copper deposits from the quartz in which they were embedded; native copper was not to be found anywhere in the western mountains. Instead, the region had comparatively low-grade deposits of sulfide ores, the kinds of ores it had taken Swansea smelter owners a century to figure out how to reduce efficiently. The only market for these western ores until the Baltimore Reduction Works was built in the 1870s was across the Atlantic Ocean, which is to say they had almost no value. However, when the Baltimore Works was built, modeled on the Swansea process, and was followed a few years later by an effort to produce silver in a similar smelting process in Colorado, a domestic demand for sulfide copper ores appeared.

By the end of the 1880s, copper production would be under way in Utah as well as Montana and would soon be followed by copper processing in Arizona, but Montana's Butte Hill and the smelters connected to it led the way. There, local capitalists and some outside investors began exploiting a high-grade copper sulfide found in some of the mines to sell to Colorado and Baltimore; as it turned out, they even found some copper glance (sulfide ores containing more than 60 percent copper), which they could ship to Swansea for a small profit. The copper developers, however, were mining for silver and gold; as far as they were concerned, they were recovering a profit only because of the valuable silver and gold contained in the ores. As was the case with the silver hard-rock miners, the iterative growth, conditioned on larger investments to contend with deeper mines and regular uncertainty, pushed them farther than they had initially intended to go and ensnared them more firmly in the elusive pursuit of profit from hard-rock mining.

One group of developers in particular, after investing hundreds of thousands of dollars in opening the Anaconda Mine, decided to make the leap to copper production, setting off a frenzy of activity across the mining district and throughout the industrializing nation. Sulfide ores could only compete if they were mass-produced. Once the Anaconda owners had determined they had enough sulfide copper ore to do so, they built the world's largest ore-processing plant and began flooding the world market with blister copper that had been refined in the West before being shipped to the East. If hard-rock mining had formalized a steady geometric growth in places of production, low-grade copper processing seemed to accelerate that growth exponentially. Investments in multimillion-dollar plants and extensive mine developments required reinvestment in the ever widening circle of goods and services needed to run the operation. In the same way volumes of gold from the region had drawn others from the East, who saw only product and knew nothing about the work required, the volumes of copper suddenly streaming east attracted eastern capitalists eager to invest in an apparently rich field but who had little knowledge of copper mining. Soon, a half dozen copper processors were churning through hundreds of tons of ore every day, and Montana became the largest copper producer in the nation at the same time copper ore lodes in Utah began to look much more promising than ever before.

The mad scramble by financial investors to keep up with investments and stay ahead of what became a rapidly falling price for copper resulted in the production of inexpensive metallic copper at a scale unseen in human history. The abundance of raw copper, in turn, encouraged an inventive episode of world-historic importance: the development and perfection of electrical delivery systems whose charge required copper wires for transmission and whose generators were filled with the metal. These self-contained power networks were made materially possible and economically affordable by the abundance of cheap copper after 1880. By the early 1890s, engineers had perfected an affordable and generally safe design and made installations in cities around the world, affordable only because copper prices continued to fall.

By the 1890s, mining copper in the West had not only changed the face of urban life forever; it had demanded economies-of-scale and cost-saving strategies of such scope that it had spawned new kinds of business firms in the West. Copper processors found that they needed to control and coordinate multiple mines, large ore-processing plants, railroads, brick-making plants, foundries, commercial businesses, timberlands, sawmills, coal seams, electrical plants, real estate, and other ancillary and subsidiary firms related to successful copper processing. These firms became more than merely *mining* companies; they each became a kind of industrial metabolism containing its own vast, internalized mineral production material economy.

Turning copper into gold, mining and processing low-grade copper ores for a profit, became the single ambition of miners and mineral investors after 1880 and gave character, texture, and momentum to the story of US metal mining in unexpected and perhaps irretrievable ways during those years.

THE BOULDER BATHOLITH

The ore lodes that became the basis for lode mining in Butte, Montana, were actually the product of two major episodes of ore deposition, followed by five others of diminishing significance. All of them occurred within a small section of a gigantic rock formation called a *batholith* (literally, rock from the depths), named the Boulder Batholith by twentieth-century geologists. The Boulder Batholith had bubbled up as cooling magma deep below the surface about 78 million years ago, prior to the onset of the Laramide Orogeny, which created the northern Rocky Mountains. This enormous igneous rock structure comprises the bedrock beneath Butte, the Deer Lodge Valley, and as far north as Helena—underlying an area roughly 5,000 square miles in extent. The Boulder Batholith is a quartz monzonite, which appears to be a pinkish granite but contains too much quartz (5 percent to 20 percent) to be classified as such.

As the Laramide Orogeny began 70 million years ago, the usual folding and buckling that accompanies mountain building affected the Boulder Batholith. In the area beneath present-day Butte, these dynamics created gigantic fissures—100 feet across in some places—that filled with superheated solution from below and quickly precipitated into large enriched quartz and sulfide deposits. Long after the first lodes had cooled and set in place, the entire structure buckled and fissured again, and a similar process created similar-sized deposits cross-cutting and bisecting the first. At least five more times over the millions of years during which the northern Rocky Mountains formed, exploding steam solution filled the openings of new fissures and deposited additional sulfide materials in this section of rock underlying the future Butte.[2]

The four-by-seven-mile piece of the Boulder Batholith comprising Butte Hill probably first became exposed to the open air about 2 million years ago as its movement skyward finally met the steady forces of climatic erosion. The fissures would have appeared at the surface as a curving series of obvious discolorations slicing across the hillside from west to east in an erratic branching pattern. To the north and east, the batholith had folded under the towering mountains of the Continental Divide; to the west, it had been broken up by a volcano; and its downward slope to the south had submerged it deep beneath the valley detritus. Most sections of the sprawling Boulder Batholith had experienced episodes of mineral vein deposition, but only a random few of these deposits both appeared at the surface *and* carried valuable metals. The deposits poking out of Butte Hill did both.[3]

Some of the Butte ore lodes were quite rich in silver, as the preliminary lode mining developments had begun to reveal. The highest concentration of silver ore had been deposited in lodes on the west side of the large group of outcroppings, nearest to the Big Butte volcano stem from which the settlement of Butte got its name. It was here on the Rainbow Lode that coursed through the northwest region of the exposed batholith that Marcus Daly had worked the silver chloride and found the substantially enriched silver sulfides in the groundwater zone for the Walker brothers' Alice Company. Just downhill and to the south of the Alice Mines, A. J. Davis had pursued similar ore in his Lexington Mine, which he worked in his Lexington Mill. If one followed these outcroppings east across Dublin Gulch, the ore lodes found there contained silver, but they also included a good deal of copper content, as William Clark was aware through his lode mining work. All of the ore lodes had been significantly weathered and oxidized over the 2 million seasons that had passed since the Boulder Batholith surfaced. On the northwest part of the mineralized zone, the outcroppings appeared as a brown or reddish quartz rock; but as Daly and Davis had proven in their claims, these unusual-looking ores could be treated mechanically or be roasted and crushed. To the east across Dublin Gulch the outcroppings were darker but otherwise similarly weathered, although their copper and iron content made them more complicated to reduce.[4]

In the silver zone of ore lodes, success had hinged in part on a miner's or a mining company's access to capital, as we saw in chapter 2; but as in placer mining, these deposits were varied and erratic rather than consistent and predictable, so location mattered. Daly and Davis had come to the Butte Hill ore lodes well-funded, but they had also chosen particularly good locations where enrichment zones continued with the depth of their mines. In other words, for all of their funding and the experience they enlisted, they had also been lucky in selecting what turned out to be rich sections of the lode for their claims. Other mines on the hill were not as rich, as Clark learned with the Asteroid Mine. Indeed, as miners began to discover through their continued development of mines in these ore lodes, the silver deposits to the north on the west side had many highly enriched pockets of silver sulfide below groundwater, but lodes just to the south in this region did not. The inconsistent deposition of silver in the northwest zone only grew more complicated as one moved into the other ores on Butte Hill. No one doubted that similar patterns of silver deposition could be found throughout these ores, but the presence of copper in the upper levels above groundwater created enough of an obstacle to reduction that none of the

Map and Inset of the Boulder Batholith

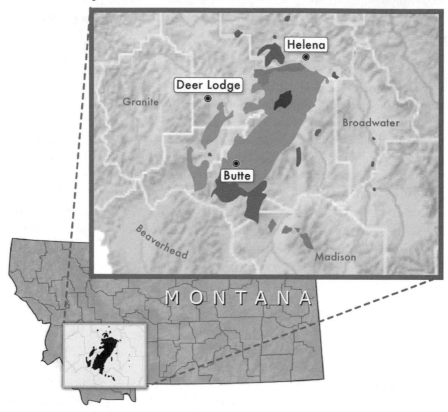

This enormous bedrock formation rested underneath a huge area of western Montana but made its most important contribution with its surfacing ore lodes in the Summit Valley, where Butte is located. Original graphic design by Sharon Hart, based on an image in David D. Alt, Roadside Geology of Montana (Missoula: Mountain Press Publishing, 1986).

resident developers could generate enough of an income from the ores to justify the greater expense of venturing below groundwater. In addition, none of the miners on the hill had been successful in finding outside investors who would pay for the development on the uncertain chance that the silver content found above would remain in the ores within the saturation zone.

The obstacle the anomalous copper presented was twofold. First, the processing techniques required for copper compounds were more expensive and involved than anything yet built in the Summit Valley, putting more demands on an uncertain mine to produce an income than developers seemed willing to risk. Second, the copper, which made up a relatively large proportion of the ore, had almost no value in comparison to silver and gold. It was valued at cents a pound in comparison to dollars per ounce for the precious metals, so it presented a significant economic problem.

On the periodic table of the elements, copper, silver, and gold reside atop one another in the same column; they are what are known as "transition metals" because of their shared pattern of electron configuration. They are also the only metals found in a pure state in nature without any need for metallurgical manipulation, so they have been used by human societies since prehistoric times. These same characteristics also help explain why the three metals were found in the same general vicinity where enriched ore lodes had been located. The electron pattern, a single electron on the outermost ring atop a completed ring of electrons, makes all of the transition metals soft; so copper too—at least in its pure elemental form—was not a useful metal because it had no hardness and could not be sharpened.

However, at some point 3,000–4,000 years ago in the Middle East, North Africa, and Europe, ancient metal workers discovered that by adding elements of tin to the copper in small proportions, a much harder and more durable metal—now called bronze—would result. In this way, copper as an alloy became the first useful metal, and most anthropologists agree that its necessary metallurgical practices forged the way toward the use of iron—which is never found in a useful state in nature—and, eventually, steel. In the same way that silver appears to have acquired a lower cultural value than gold because of its relative abundance and propensity to tarnish, copper's significantly higher abundance—almost 500 times more abundant than silver and 35,000 times more abundant than gold—as well as its higher susceptibility to tarnish, more than likely contributed to its exceptionally low economic value. While copper's abundance had undoubtedly contributed to its early use by prehistoric humans, by the modern period most of the pure or "native" copper deposited in its elemental form had been extracted for use by human society; and the remaining deposits, such as the anomalous material bound up in the ores in the central zone of Butte Hill, were compound mineral deposits in which copper was only a fraction of the overall material and tended to be chemically bound up in the minerals.[5]

Copper's low price and the complexity of its ores were further complicated by another factor that diminished its desirability: market demand for the metal was steady but generally soft, and it was easy to drive down prices by producing too much of the metal. Indeed, throughout the 1870s, professional mining engineers expressed intermittent concern that demand for copper was not keeping pace with supply. The growing supply resulted from vigorous production at the various Swansea smelters and from the regular output from native copper deposits exploited on the Keweenaw Peninsula. Some hoped new industrial uses might be found for the metal, but everyone noted that since the Civil War the decline in demand for bronze castings had precipitated a slump that might only be reversed with the return to large-scale warfare somewhere in the world. At the start of the 1870s, Rossiter Raymond himself commented during an American Institute of Mining Engineers meeting that the best use for copper was as a natural packaging for gold and silver on its way east and to Swansea. By 1878 he was still complaining: "Copper has for several years been produced so largely as to make it necessary for us to dispose of a large portion of our product in Europe."[6]

Any effort to begin exploiting the copper ores in the central zone in Butte would have to at once confront the geological tangle, economic obstacles, and the market slump, never mind the already expensive challenges of lode mining itself.

THE ANACONDA MINE

In the fall of 1880 Marcus Daly, who had left the Alice Company earlier in the year, convinced James Haggin, George Hearst, and Lloyd Tevis—who made up one of San Francisco's wealthiest financial syndicates—to invest a modest sum of money in the development of a mine Daly had recently purchased in the central zone of Butte Hill. The Anaconda Mine was among the dozens of lode mining claims that had been located on the outcroppings of Butte Hill following the passage of the 1872 mining law, and, as with many of those claims, the locators had managed to undertake only a small amount of development work in the several years since that time. The name *Anaconda* was a residual from the Civil War as the locator, Civil War veteran Michael Hickey, borrowed the name from the "Anaconda Plan," General Winfield Scott's strategy to choke the South. Either Hickey had miscalculated the capital and labor required to establish the mine, or he simply intended to hold it until an interested buyer took it off his hands. Either way, by 1880 he and his partners had only succeeded in excavating a rudimentary discovery shaft a few dozen feet deep and had begun some haphazard drifts along the face of the underground ore lode. Even with such minimal development, no one doubted that the mine held value.

Daly's purchase price of $30,000 suggested that both he and the sellers assumed it would be a silver property, perhaps similar to the Alice Mine. Daly had no shortage of experience and skill; he knew what steps were needed to further the development, but he had limited capital. Daly wasn't broke by any means; he had just sold his shares in the Alice Company for a hefty $100,000 and had $70,000 left after purchasing the Anaconda, but that would not be enough to cover all of the expenses (expected and unexpected) of mine development and ore processing. The last thing Daly wanted was to lose his small fortune on a mine investment he could not see

Marcus Daly (undated). Marcus Daly was a different sort of western miner than the men who had come to Montana with the gold rush. He had worked in the mines at the Comstock Lode and built a reputation as an expert mine manager. He was already fairly experienced when he arrived in Montana in 1876. Portrait by W. H. Hoover; courtesy, Montana Historical Society Photography Archives, catalog #941-882, Helena.

through to fruition. He needed backers with much deeper pockets than his own, so he decided to call in a favor that had been building for two decades.[7]

Daly had first met George Hearst, his entrée into the San Francisco syndicate, on the Comstock Lode in the late 1850s when both men were just beginning their western mining careers. Hearst had been much luckier than Daly in that location, but Daly had kept his old mining colleague in his sights over the years. In the late 1870s, for example, Daly had passed along information to Hearst about the Ontario silver mine in Utah, which was paying the syndicate good dividends. With his long relationship with Hearst, his good track record with the Walker brothers, and the successful tip about the Utah silver mine, Daly had provided himself with an opening to offer a deal to the San Franciscan mining capitalists. The offer to sell interest in a promising silver property in the already successful Butte silver mining district, in turn, must have delighted the men, who had no doubt watched the growing returns from the district that year with some interest. It was not the riskiest investment any of the men had made during their lucky careers in western mining, so the syndicate agreed to the offer. James Haggin began sending money to Marcus Daly to pay for preliminary development of the Anaconda Mine beginning in February 1881.[8]

No written record exists of the agreement made between Daly and the syndicate, but the pattern of behavior in 1881 provides excellent clues about the arrangements. The pattern reveals that the syndicate had agreed to fund the development of the Anaconda Mine in exchange for its share of any profits made from minerals contained within. During the winter and early spring of 1881, Daly drew enough money from Haggin's account to pay for a small team of laborers, who more than likely constructed surface buildings to enclose the mine opening—shielding it from weather—and then began to dig and frame the shaft. He drew the same amount again in early

May. At the end of May, Daly withdrew $30,000 all at once from Haggin's account, followed by $10,000 in two installments later in the summer. Daly also purchased an expensive mechanical hoist that summer. The money and machinery were engaged in what is called *dead work*, a term for the mine development work that did not directly involve removing ore. In this case, Daly's laborers worked to complete excavation and framing of the mine shaft and from there to tunnel to the face of the underground ore lode.

By midsummer Daly's miners had opened the face of the ore on several levels, spaced 100 feet apart vertically underground. Daly obviously sought to repeat his experience in the Alice Mine, where the Walkers used the valuable silver ore above the saturation zone to pay for development below the groundwater level. Displaying the careful consideration that had helped the San Francisco capitalists attain their fortunes in mining, the men would not make the decision to begin working ore or to open any more of the mine without first seeing and sampling the ore for themselves. It was one thing to hear rumors or to see ore said to have come from a particular mine. It was another thing altogether to pull that ore out of the mine yourself. They sent Hearst, the most skilled miner of the three. Ten years after Rossiter Raymond and Anton Eilers had surveyed the region and nineteen years after the first gold rush had begun drawing placer gold miners to the northern mountains, George Hearst climbed Butte Hill and descended into the Anaconda Mine to look at the ores in person and decide whether this small prospect was worth another cent of the syndicate's fortune.[9]

In what would eventually become not only the most important decision for the fortunes of all the men involved but ultimately a decision of world-historical importance, Hearst immediately encouraged his partners to pay for the removal of the ore in place and to allow Daly to continue with mine development. Daly then began drawing $10,000 a month, enough to support almost 100 miners. He purchased all of the needed equipment and supplies for the remainder of 1881 and the spring of 1882. In late spring Daly began doubling the pace of work, withdrawing cash at twice the rate to increase his labor force. He also purchased more equipment and supplies for the growing mine. The upper-level ores contained an oxidized silver chloride ore similar to that found in Clark's Asteroid claim. Daly leased the Dexter Mill from William Clark for the 1882 season, and his mill workers produced just over $200,000 in silver by the end of the year—a 33 percent profit on the syndicate's investment to date.[10]

In midsummer 1882, Daly purchased water pumps and installed them in the lower level of the mine, his "intention," the *Butte Miner* reported, "being to develop thoroughly and ascertain the character of the ore and the nature and capacity of the smelter required." Daly anticipated a repeat of what he had found in the Alice Mine

James Ben Ali Haggin (photo undated) was one of several early mining capitalists whose entire fortune had been derived from lucky mine investments. Prior to investing in Montana, Haggin had purchased early shares in the Homestake Gold Mine in the Black Hills. In this ironic twist, gold was reinvested to turn copper back into gold. Courtesy, Bancroft Library, University of California, Berkeley.

a few years earlier, with silver chloride ore replaced by a roasting ore. In fact, he was so certain about these circumstances that he placed an order for a roasting plant before his miners had opened the saturation zone. When Daly's miners finally reached the ore face and emerged from the Anaconda Mine with samples, Daly changed his plans; he canceled the order for the silver roaster and rushed to California to meet with James Haggin. The ore lode in the Anaconda Mine did not turn to silver sulfide in the saturation zone. Instead, Daly's miners opened a thick seam of highly enriched copper glance.[11]

The men had another critical decision to make. They had reinvested all of their profits in the Anaconda Mine and not found what they expected. On the one hand, finding that they held a copper mine must have been a shocking blow for a group of miners used to dealing in gold and silver. On the other hand, the mine was theirs, and the ore was unusually rich. Perhaps there was some way to turn these circumstances to their advantage, although not without taking another significant gamble and spending a great deal more money on Butte Hill—which is precisely what they decided to do. Daly returned to Butte and purchased, with what historian Michael Malone has called "surprisingly little difficulty," all of the claims surrounding the Anaconda Mine—the St. Lawrence, Neversweat, Mountain Lode, Modoc, and several smaller mining properties. The San Franciscans no doubt speculated that the nearby mines would also contain similar ore and had calculated the need for an exponentially larger supply of copper ores if they wanted to recover profits on the venture. Undoubtedly, none of the other mine owners was aware or told by Daly of their mines' potential value to the syndicate's plans—none had paid for exploration into the saturation zone—and Daly seems to have been successful in keeping his own finds and the syndicate's larger ambitions from the local mining press as well. During the second half of 1882, Daly had his miners continue to develop the Anaconda Mine and begin development on the new claims. Underground, Daly's miners performed mostly dead work to open access to the ore lodes—blasting and excavating rock, building and framing tunnels, and creating multistory underground work spaces.[12]

In late 1882 and throughout the next year, Daly began removing ores from the mines in an apparent effort to recover some of the investments, which were quickly adding up. The ore was brought to the surface and sorted. Copper deposits were no more uniform than gold or silver deposits had been; while there was a good deal of unusually rich copper glance, it was deposited in and among ores with a much lower copper content. The lower-value ore ended up in a large and steadily growing dump on the mine project. By 1884, Daly had shipped about 37,000 tons of copper glance—roughly 250 rail cars in all—to Baltimore and Swansea. The high-grade copper, worth about 15 cents a pound, sold for a total of nearly $800,000 between 1882 and 1884, providing steady operating and development funds for the mining activities.[13]

Despite the sale of so much copper glance, however, by January 1884 the syndicate was slowly losing money on its mining investments on Butte Hill. According to records assembled by Freiberg graduate James Hague, the syndicate had invested at least $2 million in its Montana developments at that time but recovered only a little more than $1 million from ore sales. The accounting did not tell a very positive story, but Hague must have known, as Daly and the syndicate clearly knew, that the accounting did not tell the entire story and perhaps not even the most significant part. Three things would have stood out as commendable expenses to Hague and his professional sensibilities. The first would have been the extensive underground work spaces constructed by Daly's miners. The ore lode had been exposed on multiple levels, not only creating significant areas in which miners could work to remove ore—thus creating the conditions for a large production of ore from each mine— but, more important, providing a higher degree of certainty about ore reserves.

Perhaps more than any other company in the history of American mining to that point, Daly and the San Franciscans had gained broad knowledge of their underground ore reserves as a first step in their mining investment. The second commendable thing would have been the ore dumps. Daly had not had his miners pile up the less valuable copper ore because he intended to discard it. Rather, he had his engineers sample and keep track of everything. By early 1884, hundreds of tons of this so-called second-class ore were stacked in enormous piles on the mining claim. Even more than exposed ore faces in the mines, these piles represented an investment in certainty. Here were reserves for which Daly had collected both weight and assays, so to the degree that complex ores could be assessed and known, these piles were

George Hearst (undated). Having made a fortune in the Comstock silver rush and another with an investment in the Homestake Gold Mine, Hearst was among the rare '49ers who made a fortune and kept it. His role in bridging the worlds of Marcus Daly and James Haggin cannot be underestimated. Courtesy, Bancroft Library, University of California, Berkeley.

legible. The third commendable development was not in the Summit Valley but twenty-six miles to the west, across the southern edges of the Deer Lodge Valley and up the gently sloping drainage of the lower Warm Springs Creek. There, to the southeast and uphill from the recently platted town of Anaconda, Haggin and Daly had employed masons, carpenters, and engineers to construct a copper smelter larger than any copper smelter ever built. In January 1884 construction was still under way in Anaconda, but the plans were clear: the San Francisco syndicate had decided to try to make a profit mining and smelting low-grade copper ores in the US West, something that had never been done.[14]

COPPER'S SYSTEM

While much has been made of the changes that would occur in Montana's mining industry in the 1880s, all too frequently attributed to Daly's "discovery" of copper ores in Butte, knowledge about the presence of copper ores in the Butte Hill ore lodes was not a secret Daly and Haggin had revealed for the first time in their Anaconda Mine. Instead, the knowledge was at least as old as the Silver Bow Mining District, where a few dozen placer gold seekers had made claims and built sluices to wash the gravels of Silver Bow Creek in the summer of 1864. At that time, several ambitious miners noted the significant outcroppings sprawling across Butte Hill. One of them, a gold seeker named William Parks, had tried in vain to reduce the blackish ore—which he was certain had to be copper—by constructing a small furnace out of brick.

In 1868, W. W. Keyes reported in J. Ross Browne's first US Treasury report that Butte's copper-enriched outcroppings were as wide as a wagon road at the places where they coursed across the hilltops above the diggings of Silver Bow Creek. Anton Eilers noted the same ores in his 1871 visit to Montana for the US Department of Treasury. Reflecting the value professional mining engineers placed on copper ores at the time, Eilers had suggested that the Butte copper ores might be used as an inexpensive and locally available flux to aid in the smelting of silver in Argenta. Indeed, William Clark had learned during his 1872 enrollment in the School of Mines at Columbia College that he held a copper mine; as soon as economic conditions improved later in the 1870s, Clark began extracting and selling the richest of his copper ores to Baltimore and Swansea smelters. In 1879 Clark convinced James Hill to help finance the building of the Colorado and Montana copper roaster in Butte. Hill's silver smelter at Black Hawk, Colorado—the first advanced smelting operation in the mountain west—needed ample supplies of enriched copper sulfide as a flux for the processing of silver ore (Eiler's initial suggestion was not far from the mark). The Butte facility would be capable of processing Butte's ores into a standard-grade copper matte for special shipment to Colorado to make silver.[15]

By 1882 four copper smelters, including Clark and Hill's, had been built near Butte Hill—all of them designed to convert the medium-grade copper ore mined from under the region into a high-grade copper matte. The Hecla Company in Glendale, which had earlier constructed a very large silver-roasting mill with the hope of capturing the Butte custom silver ore market, installed a Swansea-style smelter shortly after Clark in 1879 to treat the copper ore piling up in the district's mine dumps. The following year Charles Meader, who had come to Butte on behalf of Boston capitalists, began constructing a smelter and platting the new town of Meaderville northeast of the city. When he saw the opportunities that still existed in the mining district, Meader quickly abandoned his initial project and set out on his own, purchasing the Bell Mine and constructing a smelter to treat the rich silver-copper ores found in that mine. A. J. Davis—who had enjoyed some lode mining success with his Lexington Mine and Mill—joined Helena capitalist, former vigilante, and failed silver miner Samuel Hauser and two additional Connecticut investors to purchase the Parrot Mine (the oldest copper mine in the district). By 1881 they had constructed the Parrot Smelter and installed customized Bessemer converters (usually used for steelmaking) to work the mine's copper-silver ores. All of Butte's new smelters were designed to convert rich copper ores into either an even richer copper matte—a semi-refined metallic sulfide containing upward of 70 percent copper—or a nearly pure blister copper. Both were then shipped east for further processing or for sale to copper fabricators.[16]

Daly had watched these developments take shape, often at the hands of hired trained professionals and experts, as he made the transition from the Alice to the Anaconda. In each case the smelters were designed to work twenty to forty tons of ore a day, a larger amount of ore than any of the silver mines or mills had approached but one that was economically necessary because of the lower price paid for copper. Each of the four copper smelters built by 1882 used the same general processes designed to convert the raw ore into more industrially useful copper matte. The raw ore was first roasted in huge open-air stalls in which alternating pallets of dried timber were placed atop thick piles of raw ore, and the whole structure was lit on fire to burn for a week or more. The goal was to burn off a significant portion of the sulfide making up the majority of the ore. This roasted ore was then treated by fire, first in a reverberatory and then in a calcining furnace of some kind to separate more sulfur and, through high heat, to precipitate iron compounds and other impurities out of the molten material before it was cast into shiny ingots of higher-value concentrated copper material. Each of the individual smelting companies approached the reduction problem on the basis of the specific ores found in abundance in its individual mine holdings, so each had a slightly different process and different practices in place, but they all used some combination of these practices to achieve a generally uniform copper matte or blister copper.

When Daly's miners had opened the saturation zone in the Anaconda in 1882, he had the evidence of the four copper-processing plants to show him that a high-value copper product could be produced in the region from these ores. His move to consolidate all of the nearby mining claims, a form of horizontal integration designed to reduce market competition, reflected his guess that they all likely also held ore lodes with equally enriched copper ores; but as he came to realize the generally limited deposits of copper glance as well as the extensive deposits of the region's well-known "smelting ores," his ultimate plans must have begun to crystallize. By 1883 Daly's miners had excavated enough lower-grade copper ores and exposed enough of the underground lodes to feel some certainty that the mines contained a significant volume of copper ore—more ore in sight than any of the existing copper companies had—but their effort to overcome material uncertainty in the mines raised new uncertainties at the surface.

The cost to build and equip a processing plant represented the single highest expense a copper producer faced, and mineral processing—the crushing, grinding, melting, and casting of material—was almost as destructive to the plant's equipment as mining was to the value of the ore left in the ground: mineral processing ground through capital. If Daly and the San Franciscans were going to make a return on their investment in the mine's development, they would need to make a very careful decision about the scale of the smelter they were going to build. They knew they had a large reserve of ore, so building a plant as small as the others in the valley seemed unwise, as the equipment would surely wear out before they had recovered their costs. On the other hand, building too large a plant created several kinds of risks.

There was a chance that producing too much copper for an already glutted market could drive prices too low to afford operation. There was also the possibility, especially if prices dropped precipitously, that they could run out of smelting ores before recovering their costs. Finally, there was the issue of limited space and Butte residents' growing discomfort with the thickening air pollution caused by the roasting and smelting already taking place in the valley. This made Butte a high-risk location in terms of access to water and allowing effluent, never mind the ongoing disposal space for tailings and slag—the major solid-waste products of copper smelting.[17]

With these risks in mind, Daly had his men begin scouting locations for a smelter somewhere outside the Summit Valley. They were seeking a particular geography, a location that would have relatively easy and inexpensive access to the syndicate's mine complex in Butte, an ample year-round source of flowing water, and sufficient land on which to build the processing plant itself, surrounded by additional sufficient land on which to dispose of the tailings and slag. After an extensive search and eliminating several promising sites, the partners decided on a small finger valley drainage spilling into the southwestern end of the Deer Lodge Valley, twenty-six miles west of and downhill from Butte, over an easy grade (incidentally, about twice that distance south and upriver from Gold Creek at the north end of the Deer Lodge Valley). Here, Warm Springs Creek, whose "tributaries [were] sheltered by forests that protect[ed it] from flood and drouth," according to the *New North-West* (published in Deer Lodge City, Montana), supplied enough water to support a smelting plant. Fortunately for Daly and his investors, other than the occasional wandering herd of cattle, the flat, sloping drainage had no occupants and was available for sale.[18]

Daly had his purchasing agent, attorney Morgan Evans, begin acquiring the necessary land for the copper-processing facility in mid-1883, more than a year after first finding copper glance in the Anaconda Mine. While it was estimated that the plant and its tailings dumps would require approximately 10 acres, Evans purchased more than 9,000 acres—nearly the entire valley floor of the Warm Springs Creek drainage as well as the slopes on either side—giving the company ownership of a huge contiguous tract surrounded on three sides by mountains and hills. The *New North-West* speculated about the motive behind this large land purchase. "These smelters in a few years destroy all vegetation in their immediate vicinity," the newspaper reported after the sale was announced. "If the company owns all the cultivable land to be injured, no damage can occur to any other person by reason of the smelters." Daly and Haggin certainly wanted political control over the landscape surrounding the new facility and carved a third of it—3,000 acres in all—to be platted and incorporated as the city of Anaconda, Montana, the settlement where most of the smelter workers would live and whose sale of house lots would provide additional real estate income for the company. The remaining lands would be held by the company. It would build its processing plant on some of the lands, with ample room for expansion, additional space for storage or other unexpected land needs, and a timbered hillside.[19]

Once the plans for the Anaconda Smelter were made clear, everyone understood why the syndicate and Daly had looked beyond the Summit Valley. They had huge ambitions. They intended to construct the largest ore-processing facility ever built, capable of grinding through more than 400 tons of ore a day (ten times the capacity of Butte's largest plants at the time). Daly and Haggin intended not merely to compete with the other copper producers in Butte or merely to produce copper matte as a flux for silver smelting elsewhere in the West; they had their sights set on the national copper market, which Lake Superior's native coppers had dominated since the late 1840s. They aimed to process enough ores at a low enough cost that they could weather an inevitable price collapse and still remain economically viable. To do so, they would have to mass-produce their copper matte on a scale unprecedented in the history of copper production—achieving, they hoped, a cost savings that would allow a second-class copper sulfate ore to compete with the native copper from Michigan that was much closer to eastern and European markets and did not require smelting. They had set quite a challenge for themselves.[20]

Seeking every opportunity to cut costs and knowing the tremendous steady demand such a smelter would make on the Anaconda Mine and, in turn, on timber requirements, Daly formed a partnership with Edward Bonner and three other Montana capitalists to create the Montana Improvement Company (MIC). The company would organize timber acquisition and harvest and produce mine and smelter lumber at a large mill west of Anaconda. The MIC also maintained a surplus stockyard in Butte that sold mining lumber to the other mining interests in the valley.[21]

Daly and the San Franciscans, on the basis of their known ore reserves and with the economic challenges they knew they were about to face in bringing their copper to market, had done more than simply become another lode mining firm in the West. They had built the foundation for a copper production system, and they had done so carefully and with a keen eye toward the future. Although none of the men had trained in Europe, they appear to have absorbed the cautionary lessons presented by the professional mining engineers during the 1870s in their careful and rigorous planning, and they went to some extent to hire professionally trained men and to copy known European processes. Only *after* access to sufficient necessary resources—ores, water, land, and timber—had been secured did construction begin on the smelter in Anaconda.

On August 11, 1883, the *Engineering and Mining Journal* reported that "excavation for the new smelter [in Anaconda] began July 23rd." The largest part of the smelting works would be its massive ore concentrator, an enormous wood building built into the side of the hill northeast of the new platted town, designed to use gravity to drop the ore through crushers and screens that would separate the metallic ore from the waste material and rock and to crush the ore into a uniform size for transport to the roasting and calcining plants—uniformity increased the efficiency of the heat that would be needed to reduce the ores in later processes. When completed, this concentrator would be able to crush and agitate its way through more than 400 tons of raw ore a day. The roasting and smelting buildings were equally impressive, with dozens of large roasting and reverberatory furnaces designed to transform mass amounts of the copper concentrate to a copper matte averaging 60 percent copper. Most of the plant's individual smelting techniques, as well as the general flow of the ore-processing system, had been developed elsewhere (most of them in Swansea, Wales, where Daly had sent his engineer Otto Stahlmann to study smelter design in 1883), but the arrangement of so much processing capacity in one place was unique to the Anaconda works; its scale was unprecedented.

During 1883 and 1884 Daly employed several hundred men to construct the facility, which cost slightly more than $2 million, including the purchase of land, water rights, construction supplies, labor, and machinery. The first Anaconda smelter would take more than a year to build, but it would represent a landmark achievement in natural resource production and exploitation, as well as an innovative feat of industrial system design. As the smelter was nearing completion in summer of 1884, the Utah and Northern Railroad, which had arrived in Butte two years earlier, constructed a spur line from Butte—called the Montana Union—that connected the Anaconda Mines to the Anaconda smelters by rail. On September 7, 1884, the Montana Union hauled the first trainload of ore west from the Anaconda Mine and dumped it into the top of the concentrator at the Anaconda Smelter. A new era in copper production had arrived.[22]

The startup of the Anaconda ore-processing facility in September 1884 sent ripples through the national and international copper markets. By itself, the production from Anaconda's new smelter doubled the ore-processing capacity of the Butte copper mining region, which had already pushed prices down and exceeded the capacity of the Baltimore Copper Refinery. The new plant contributed so much copper to the national market that copper prices plummeted. The copper mills in the Lake Superior District increased their production in response, setting off a price war and sending copper prices down even further. Daly and Haggin had anticipated this response and kept up a steady pace of mining and processing. By 1885 Daly's miners were extracting one-quarter more smelting ore than his smelter could process each day, building up another significant stockpile of sampled ore. In the East, producers had clearly not anticipated Daly's holdings or his gumption. The Baltimore

Anaconda, Montana, 1887. The first Anaconda smelter was built in 1883–84. This wooden structure contained a scale of mass production throughput unlike any smelting facility before it, capable of processing 400 tons of ore a day. Within a few short years, this plant proved to be too small. Photo by Hazeltine.

Copper Refinery, the best-developed Swansea-style smelter in the country, found itself scrambling to expand to accommodate all of the copper matte suddenly produced in Montana.[23]

As the price of copper continued to fall and the stockpile of smelting ore in Butte grew larger by the day, Daly and Haggin began plans to achieve even greater economies of scale to reduce their costs further in a falling market. The men developed plans to fully overhaul their ore-processing plant and to double its capacity in the process. Daly had begun to develop the St. Lawrence, Neversweat, and other mines surrounding the Anaconda after the Anaconda plant had gone into operation. His miners developed the same extensive tunneling on multiple levels to the ore face that had made the Anaconda Mine so productive. Below the water line, these mines held somewhat different copper compounds and were richer in silver than the Anaconda ores. Daly and Haggin decided to construct a second ore-processing plant designed specifically for the ores from these other mines. In 1886 they had workers begin construction on the expansion of the first plant—soon called the Upper Works—to

Anaconda Mine, smelters, 1885. Courtesy, University of Washington Libraries Special Collections, Seattle.

accommodate the large ore production from the Anaconda Mine, at the same time other workers began construction of the Lower Works, a smelter designed for the ore compounds found in the other mines. The goal was to achieve a combined ore-processing capacity of 4,000 tons a day, ten times the capacity of the original plant three years earlier.[24]

Despite their adherence to the careful planning, preparation, and systematic development advocated by the professional engineers, Daly and the syndicate found themselves growing and expanding even faster than the silver processors had in the late 1870s. In the first case, developing the certainty they needed to justify a smelter—which meant literally developing the underground by excavating mines—raised their overall investment such that accomplishing a return on investment would require a large processing plant. A large plant could not be safely located in Butte. The choice to invest another $2 million in the world's largest ore-processing plant and to develop an entire community made sense on the basis of known ore reserves and a desire to find additional means of generating income (real estate has always been a very good investment in the United States), but its consumption of ore resources and operating costs combined with the steadily falling price of copper further accelerated the expansive tendencies. On the one hand, a decreasing quality of ore as the men smelted an average grade that became more and more difficult to attain very quickly rendered the original processing plant too small and inefficient for their goal of turning a profit. In a related manner, the next two mines to come under development—the St. Lawrence and the Neversweat—had a slightly different mix of minerals in their ore and required the construction of a second processing plant to be most profitably treated. Once again, major investments of capital and labor did not create a "stately and enduring edifice," as Rossiter Raymond had promised in 1869; rather, they generated ongoing material revisions that, in turn, encouraged an iterative increase in scale to gain the best economic advantage under the circumstances.[25]

The increases in scale were not limited to intensifying mining and enlarging smelting. As was the case with the earlier silver-processing business, they seemed to ripple out into the surrounding markets, although this time the mineral processors did not sit idly by and wait. F. E. Sargeant, an early manager for Daly and Haggin, described periods of scarcity and delay while constructing the two enormous copper processors in the late 1880s. Rather than wait for others to deliver needed supplies from as far away as San Francisco, Chicago, and New York, the men decided to build up their own stock of construction supplies. "This resulted in the organization of what was known as the Anaconda Commercial Company," Sargeant wrote, "and another organization was also formed at the same time in Butte, to furnish material to the mines, called the Butte Trading Company. Both of these companies furnished nearly all the material used for several years and effected a great saving in building and running the works, and in operating the mines." When the smelter began to work the ores, the wear and tear on the machines produced a growing demand for parts that also could not afford to wait for the timetable of others to catch up: "To meet this want a foundry and machine shop was started which has grown with the company's growth, kept pace with its rapidly increasing needs, and finally consolidated with the Anaconda Commercial Company and the Butte Trading Company, forming the Tuttle Mfg. & Supply Co., with its foundry, machine shops and principal supply store at Anaconda, and a branch store at Butte." Sargeant credited this vertical integration with a critical role in lowering costs of production: "This organization has protected the Anaconda Company from paying excessive prices for supplies, has kept the fires of its furnaces going night and day to prevent loss of time at the Smelters, and has been a very important factor in the successful and continuous operation of the works."[26]

Daly and Haggin had also made capital investments that went well beyond mining and its technical and resource necessities. They created the Anaconda Town Site Company to sell house and business lots, generating a profit from the land purchased in the Warm Springs drainage. They created the Montana Hotel Company, which constructed an enormous brick hotel in the platted town of Anaconda, towering five stories above the open valley floor long before the town itself had been filled in around it. Daly also opened a racetrack. In less than a decade, the careful development of a presumed silver claim had ballooned into a sprawling enterprise of mining, milling, smelting, manufacturing, racing, real estate, and retail sales. The dynamics of bigness seemed to defy the developers' efforts at containment.[27]

The Anaconda properties and businesses represented only the most extensive expression of a tendency prevalent throughout the Butte copper industry during the 1880s. By 1886 the Montana Copper Company, which had begun running the Calusa Smelter in 1881 with an initial capacity of 10 tons of ore a day, had rebuilt and expanded the smelter to fifteen times its original size, with a capacity of 150 tons a day. The Parrot Smelter, operated by Boston financiers who hired Freiberg graduate and copper expert Edward D. Peters to improve their plant, began their concentrator in 1881 at 12 tons a day; by 1886 the smelter was twice as big as the Montana Copper Company, with a concentrator large enough to process 300 tons a day. The Colorado and Montana, Clark and Hill's collaboration, also grew from 10 to 60 tons a day in 1884 and to 100 tons a day in 1886. Highlighting the changing face of mining in the Summit Valley, historian Fred Quivik made this comparison: "After 1882, there was little growth in the capacity of Butte's silver mills. Growth in the capacity of reduction works for copper, however, was phenomenal."[28]

The pattern of accelerated iterative expansion seemed rooted in the very dynamic of western copper exploitation. The Anaconda Mine and Smelter developments drove down copper prices, making the other companies' ore less valuable at the same time. In 1882, copper had traded at just over 18 cents a pound; when the Anaconda smelters went into operation in 1884, the price fell to 13.25 cents a pound and to just over 10 cents a pound by 1886. Copper would not trade again for more than 15 cents a pound until after 1900, when consolidation of the Butte mines would slow down production at the same time demand would finally accelerate ahead of supply. The falling price had encouraged an increase in mine development, ore removal, and ore processing. As these companies dug deeper into the earth, their copper ores diminished in quality, falling to 10 percent or less copper content and creating another change in conditions that affected overall cost of production. The lower-value ore needed to be processed in larger batches and at a faster rate to stay ahead of investment. The exponential growth in ore processing during the 1880s had begun to significantly burden the air quality of the Summit Valley, filled with more miners and others than ever before. As Daly and Haggin had predicted, this growth had reached the limits of disposal lands along Silver Bow Creek, where most of Butte's copper-processing plants had been located. By 1889 six good-sized copper smelters crowded the Butte mining settlement. In contrast, the Anaconda ore-processing plants, with their seemingly unlimited access to disposal lands in the Warm Springs Valley, had avoided such constraints.[29]

In the winter of 1889 James Hague—a Freiberg-trained mining engineer, participant in the 1872 King survey, and long-established national mining consultant— completed a business report on the Anaconda properties and mailed it to James Haggin in San Francisco. Hague had spent several weeks during October and November 1889 assembling a record and an accounting of the Anaconda properties for a possible sale. Hague's assessment was quite positive. In his closing to a very long letter detailing the general finds of his report, Hague commended Haggin for his shrewd investment, "not in one mine alone, but in several well-developed and productive mines." These same mines had been managed by Daly since 1882, and Hague made special note of the fact that "the mines are opened far in advance of their immediate needs, and their productive capacity is assured for years to come."[30]

In an interesting structural parallel to the gold rush, the overall success of copper mining appeared different depending on the scale examined. Similar to the gold seekers who scrambled to every opportunity because of the radical material uncertainties on which their ambitions rested, the copper industry in Butte scrambled to keep up with unexpected and unpredicted conditions under which its perceived sure thing might find the actual financial return slipping farther into the future. From the investors' perspective, there was a certain level of chaos in the risks they had just taken. When one assessed the Butte mining region as a whole, however, one salient and seemingly triumphant fact stood out: this region had suddenly, with a speed and growth unprecedented in the history of human society, become the most important copper-producing region in the world. It had also, arguably, changed the copper industry forever. Just as the gold rush period had social impacts well beyond the gold seekers' ambitions, the system of copper processing created ripples and residues that extended far beyond its investors' fiduciary ambitions.

COPPER AND ELECTRIFICATION

Few technologies of the modern world have become more symbolic than domestic electricity, often represented by the incandescent light bulb but experienced as the proliferation of electrical-generation networks in urban centers in Europe, the United States, and Japan and the gradual spread to all modern settlements. In the twenty-first century, electrification has become culturally naturalized to such a degree that its absence becomes a noteworthy, almost unnatural event; but even though it is a natural physical phenomenon, electricity had to be produced to be used for domestic power. Our iconic story about the control of electricity usually begins with Thomas Edison's perfection of the incandescent electric light bulb in 1880 and his successful electrical distribution business, the Pearl Street Station, in lower Manhattan in late 1884. But it is more accurate to say that Edison invented the industrial laboratory and made invention—a folk practice that had previously served craftsmen and farmers alike—a commodity all its own.[31]

This iconic story of individual invention has its counterpart in the history of copper mining, where western historians have presented the common-sense idea that the exponential acceleration in copper output was caused by the demand for copper generated by the invention of electrical systems and the urban domestication of electricity. Michael Malone's *Battle for Butte* points to the telegraph, telephone, and incandescent lamp as the sequential and overlapping causes of a mushrooming demand for copper in the "new age of electricity" brought on by these inventions.[32] William Cronon's "Kennecott Journey," which explains the urbanization of the distant copper resource in Alaska, likewise argues that copper ore was considered important during an 1885 expedition because of "copper's ability to conduct electricity."[33] Arizona as well, it is said, responded to the pull of growing markets for electricity in the 1880s. Across the mountain west, it seemed, electrical demand led to the development and growth of the copper industry.

This is an easy conclusion to draw because copper has what one might call a natural suitability for electrical power. The only element more conductive—or less resistant to an electrical charge—than copper is silver, which is far too expensive to be used in the quantities needed by an electrical system. Copper's low resistance is derived from its atomic form. Copper atoms form a matrix on the molecular level, where positively charged nuclei share their electrons in a field of evenly spaced electrons—"sheets" of nuclei resting on fields of electrons, layer after layer. The copper atom's single free electron forms into electron fields, which instantly become fully charged with an electric current without losing much force. Copper's unique suitability for large-scale electrical distribution systems means it was used in unprecedented volumes when such systems came to be built. One of Edison's prototype electrical systems, for example—prior to and smaller than the Pearl Street Station—was estimated to require almost $60,000 in copper, which at 1880 prices amounted to nearly 280,000 pounds of the metal. Edison's dynamos alone needed more than a ton of copper each. The relative abundance of copper in the earth's mantle not only contributed to copper's lower price, it also assured that, if copper could be extracted and produced, there would be enough of the metal to build large-scale urban electrification systems. Electrical distribution systems, in turn, require more copper than anything before them had.[34]

Without discounting Edison's unique and powerful abilities, a closer look at the relationship between electricity and copper production raises questions about cause and effect during this important period of technological change. For example, if the demand for electricity caused the growth in copper output, one would expect that copper output would have remained relatively flat until the 1880s and that, as supply caught up with demand, prices would rise. In reality, after the Civil War, when engineering solutions designed by James Hague solved the challenges of the native copper ore mines on the Keweenaw Peninsula and began flooding the American market with copper, copper output began to increase significantly and prices started to fall. According to historian Charles Hyde, "During the quarter century that followed the Civil War, Michigan's copper output grew explosively . . . generating copper prices too low to allow most producers to earn profits."[35] Because of the Michigan industry, US copper output had doubled by the late 1870s. Output had doubled again by 1882, as railroad development reached the copper ore districts of Montana and Arizona, connecting them to the Atlantic and Pacific markets. US copper output more than doubled again by 1887, again by 1893, and yet again by 1900.[36]

Growth in copper output took off during the late 1870s, but electrification did not begin to consume a significant share of the copper market until the mid-1890s. Prior to those years, electrification was still in development. Edison had perfected the incandescent light bulb in 1878 and constructed a rudimentary and rather dangerous electrical distribution system in lower Manhattan in 1882. But despite Edison's best efforts, neither invention led directly to the immediate or widespread adoption of electricity as the illumination or power source of choice. Edison's direct current system was not only dangerous; it also had spatial limits, leading it eventually to give way to Nikola Tesla's less dangerous and less limited alternating current distribution system, perfected in the late 1880s. Not until urban streetcar traction systems became widespread in the 1890s did electrical systems of any kind influence copper production; and it was not until the turn of the twentieth century that electrical distribution systems consumed a controlling share of the nation's copper output, at which point prices began to stabilize.[37]

The influence of electrification can be seen in the refining methods copper producers used. Copper is an excellent conductor in its pure atomic state, a condition rarely found in nature and entirely absent in the western copper districts, which were producing nearly half of the nation's copper by the mid-1880s. When even small amounts of so-called impurities are present (trace metals like arsenic and antimony, which were common among the western ores), conductivity decreases significantly. The only way to ensure an "electrical" copper from these ores is to purify the copper product in an electrolytic bath, where an electrical charge is run through sulfuric acid in which copper plates are submerged. During more than twenty years of accelerating copper output, electrolytic copper represented a very small share of overall copper product; it was an expensive specialty market until the 1890s. Electrolytic

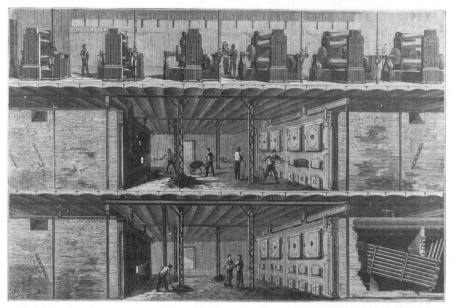

Inside Pearl Street Station, circa 1883, where laborers shoveled coal into furnaces below, making the steam to turn the copper-laden generators and distribute electricity to the homes of dozens of New Yorkers. Courtesy, Smithsonian Institution.

copper refining did not begin in earnest in the US copper industry until about that time, when the western industry suddenly began to process larger and larger portions of its copper output through new electrolytic refineries installed at the smelters. In other words, western copper output accelerated for twenty-five years before producing the specific commodity needed by electrical-generation systems.[38]

If we shift our focus from the urban centers of invention to the rural centers of production, it is possible to see that the iterative patterns of growth experienced in Montana were present for the same reasons in Michigan's original copper production region. There, the "native" coppers were among the first copper lode deposits in the country to receive professional attention when James Hague was brought in to help just prior to the Civil War. Copper output grew in Michigan to pay for the heavy capital investment necessary to process the ores, and it could only barely do so, flooding the markets and driving down copper prices all the while. Initial copper developments in Montana and Arizona followed a similar pattern. The major impetus for rapid growth in the Butte Mining District during the 1880s was not the incessant pull of market demand but rather the cutthroat tactics of trying to produce one's competition out of existence. The "price wars" of the 1880s—a strategy to flood the market with copper to reduce the price low enough to drive competitors out of business, with the hope that this happened before the company drove itself out of business—were likewise not responses to market demand but were precisely the opposite. They engaged a tactic of taking advantage of a *limited* market. Growth in output was part of a fight for market share, not a response to a market need.[39]

For copper producers, the overproduction and consequent falling prices were necessary (if undesirable) difficulties Daly and the others were determined to face in their pursuit of market share, but for the broader Atlantic marketplace, this overproduction was experienced as a growing surplus of increasingly inexpensive copper. Historians agree that Thomas Edison's incandescent lighting system project represented, more than anything, Edison's ability to fully grasp the systemic context of the inventions he pursued. According to historian Thomas Hughes, Edison's inventive process began by identifying particular kinds of problems for which technology, funding, and other "factors" appeared favorable. In other words, Edison chose problems with likely solutions. Edison "conceived of the problem to be solved by invention as inseparably technical and economic," Hughes wrote. "He did not set out to invent a lighting system the cost of which would not be considered until it was built."[40] Historian and Edison biographer Paul Israel agrees, emphasizing Edison's profoundly deductive and synthetic imagination as well as his meticulous attention to context, particularly material and economic contexts. Israel argues that Edison always undertook a characteristic "exhaustive search for plentiful supplies" prior to beginning any creative initiative. He concluded that Edison's careful search for resources was "guided by an underlying worldview that associated the bounty of nature with human technological and social progress."[41]

The growing abundance of increasingly inexpensive copper finding its way to merchants on the East Coast would have appeared as a representative case of nature's bounty, as Edison ensured when he first got started on the problems associated with inventing an incandescent lighting system. The first of Edison's Menlo Park notebooks concerns itself with copper, containing tables calculating the conductivity of variously sized copper wire. Edison had his scientific associate, Francis Upton, calculate dozens of scenarios to allow a quick estimation of the rate of copper needs and costs as he worked out his system's material demands. They reveal the importance of the cost of copper for his electrical system and highlight the importance of the metal's falling price since 1867.[42]

Indoor illumination systems had already been invented by the late 1870s, with gas lighting a popular way to light up the nighttime indoors and arc lighting a favorite streetlight design. Edison would have to develop a system less dangerous than arc lighting and less expensive than gas if he hoped for a general adoption of the incandescent technique. As it turned out, copper prices held the key. By the late 1870s,

Pearl Street Station coverage map. Edison's 1884 electrical delivery system was capable of bringing a direct current of electricity to a small section of lower Manhattan centered on Pearl Street, where his generators were housed. The system was limited in range and turned out to be rather dangerous. Courtesy, Smithsonian Institution.

between the steady growth of output in Michigan and the rapidly growing copper output from Montana, these prices dipped low enough for Edison to justify beginning system development. He must have been even more delighted to watch supplies grow and prices plummet throughout the 1880s.[43]

But not even Edison could have predicted the explosive growth in electrification that would take place by the late 1890s; nor could the Montana, Arizona, or Michigan copper producers have anticipated the vast market for copper widespread electrification would create as the twentieth century began. Electrical demand did not cause the explosive growth in copper production. Instead, it is more apparent that the production of both copper and electrical systems established a co-evolutionary trajectory, with neither possible without the other. Copper's overproduction until the late 1890s is not, however, tied to the expansion of electrification but instead stems from the internal dynamics of copper ore processing itself. The steady overproduction of copper among western producers, their Sisyphean efforts to produce ahead of falling prices and to recover the mounting costs that came with such increases, flooded the world copper market with more raw material at a lower price than had ever existed. In the context of such an abundance of material—which, because it was derived from a natural resource, had the aura and feel of being natural—the development and construction of widespread electrification became both imaginable and possible. This is not to say that the expansive tendencies of the copper production system caused the electrical revolution in modern life; far from it. Rather, it is to make a case that electrification did not cause the expansion of copper mining, as mining historians have long claimed. It is also to recognize that the overproduction of copper, driven by a dynamic inherent in US-style lode mining, produced material and economic conditions within which electrification could grow and flourish. The fact that, except during World War I, copper prices never came close to the mid-nineteenth-century high during the entire early period of electrical system construction suggests that production continued to stay on pace with consumption throughout the period.

CONSOLIDATION

As encouraging as the overproduction of copper was for the rise of urban electrification, it generated an unstable business dynamic within the Butte copper industry and a persistent instability in world copper markets. In the late 1880s, efforts were made to stabilize the market. In what became a global scandal and led to the imprisonment of several of its architects, a French syndicate tried to corner the copper market, with the aim of buying up world supplies and then withholding copper from sale to drive up prices. The European-based Rothschild family, whose gold-mining-derived fortune made them important global financiers by the late nineteenth century, saw a similar opportunity, investing significantly in copper with the same aim. There was even talk among the producers of voluntarily holding back on production, but the economic dynamics of copper production made such a move unthinkable. In Butte and elsewhere in the copper production regions of the West, mining and smelting ground on at an ever expanding rate.[44]

For Daly and Haggin's sprawling enterprise, this meant continued expansion in multiple directions, formal incorporation, and eventual sale. The men had been set back significantly in 1889, first when their completed Lower Works ore-processing plant—constructed entirely of wood—burned to the ground shortly after opening and then when the extensive timbering that kept the Anaconda and St. Lawrence Mines open caught fire and burned for months. The men expanded timber operations into the Bitterroot Valley, south of Missoula, and opened a sawmill there. They also invested in their own railroad, the Anaconda and Pacific Railroad, which ran from the mines in Butte to the smelters in Anaconda; and they opened an electrical-generation plant to supply electrical power to both smelters and to the town of Anaconda. They seemed to have had an interest in cashing out on their investments in 1889, as they hired James Hague to develop an accounting of their costs and expenses since 1882, but the numbers did not entirely favor them. In any case, no one made an offer to buy.[45]

By the mid-1890s, after yet another crippling economic depression, the men decided to incorporate, creating the Anaconda Copper Company in 1895. The company's list of business activities filled nearly a full page of text. The company held and worked more than two dozen mines on Butte Hill and was now working its deepest mines more than 1,500 feet underground, each mine representing the equivalent of an inverted 100-story building. It provided for the mining and smelting needs through timber and lumber operations, an industrial foundry, a brick-making plant, a large wholesale importer of mining equipment, coal mines, and a small regional railroad. Daly and Haggin controlled water and land and real estate in Butte and Anaconda beyond their copper production needs and managed a vast enterprise that reached into all corners of western Montana.

In 1896, according to another round of accounting performed by James Hague, the company had generated a profit of $4 million during its first full year of incorporated activity. This outcome—which resulted largely from the formalized Anaconda Copper Company's ability to begin with balanced books that artificially smoothed over costs and expenses that had gone into constructing (and reconstructing) the company's sprawling edifice—accomplished what the men had been seeking as early as 1889; they sold large shares of the business to men of finance who could afford the enormous operating costs and wait patiently for their return on investment. Significant shares of the Anaconda Copper Company were purchased by the Rothschild family, who seem to have drawn a regular profit from their investment in the standing company. By the end of the decade, however, the combination of a continued price decline, a precipitously falling reserve of smelter-grade ore, and processing plants whose age was showing led to another round of restructuring such that by 1899 the Anaconda Copper Company had become the largest single business component of one of the nation's largest trusts—Amalgamated Trust—led by Henry Rodgers and William Rockefeller. They intended to stabilize the operational growth in Butte by, one by one, buying out or out-competing the other large firms mining in the valley.[46]

Throughout the busy 1890s, as larger, more capital-intensive companies grew even larger and more capital-intensive, they also excavated more ore at a faster rate. This pushed their mines deeper and deeper into the earth in pursuit of what they hoped would be a continued supply of copper ore lodes embedded in the Boulder Batholith exposed on Butte Hill. These depths soon precipitated conflicts about the ownership of ore deep underground based on its relationship to the surface outcropping, as legislated by the 1872 mining law. "The vein system of the district is intricate," mine engineer and historian Thomas Rickard wrote about Butte, "and the dislocations due to the faults have greatly increased the complexity of structure underground, creating conditions that completely stultify the so-called law of the apex." The fissures in the batholith had formed erratically in response to heat and pressure; they had also been fractured and fragmented by secondary fissures five times in all, creating additional erratic ore deposits with each fracture. Complicated and erratic ore deposits were exacerbated by complicated and erratic surface claims, which overlapped in some places and left unclaimed pockets in others.[47]

Given the complicated shape of the ore lodes and the fact that miners were tracking them through solid rock, it was inevitable that the day would come when miners in one claim would hear the distinctive drilling and muffled sound waves of dynamite directly below them in the earth, indicating that another company was mining beneath their claim. When this day came and mine owners were notified, it became practice—despite the provisions of the 1872 law—for the claim owners to file a trespass lawsuit against the company at work under their claim, asking the court to prevent any further mining unless that company could prove ownership. The 1872 law gave miners the right to all ore lodes that showed an outcropping within the boundaries of their claim—called an "apex" right—and granted them the right to follow their ore lode as far into the ground as it went, even if doing so caused them to mine beneath the ground of another adjoining claim. The law did not explain to judges how such cases were to be decided, but in an early lawsuit in Nevada, Ninth Circuit judge Thomas Hawley set the precedent that if the claimant could show that ore was being removed from beneath his mining claim, "the party taking it out, if he does not own the surface, must show that he has a right to enter upon that ground and take out the ore."[48] In other words, if a mining company followed its ore so deep underground that it would be mining beneath another's property, it had better be prepared to prove its right to do so.

The evidentiary challenge for winning such cases in the Boulder Batholith at Butte was first made plain in a trespass suit a small, nearly undeveloped claim named the Little Darling brought against the Blue Bird mining company, whose well-financed mine was 600 feet into the ground in 1889. Little Darling's owners were successful in convincing the court that the Blue Bird miners were working in ground beneath their claim, forcing an injunction on work in the Blue Bird Mine and forcing Blue Bird owner Ferdinand Van Zandt to mount a defense. Both sides hired professional mining engineers, with the owner of the Little Darling hiring David W. Brunton and Van Zandt retaining none other than Rossiter Raymond and James Hague, whose ambitions to lead the nation's mineral development in the 1870s had slowly faded along with the profession, becoming part of a collection of commodity capital and knowledge that helped large mining companies function in the late 1880s.[49]

Brunton recalled the complexity and uncertainty that muddied a clear-cut case according to the standards of the 1872 law: "The situation proved extremely complicated and one that was never contemplated by the framers of the Apex Law." According to Brunton, the vein in question was a "true fissure vein" that had been dislocated by a fault "to such a degree that it resulted in two parallel outcroppings of the same vein in adjoining mining claims." Hague, who had simply provided an affidavit on behalf of the Blue Bird because he had been in Butte on Daly and Haggin's behalf, declared that the ores found at the 400-foot level in both the Little Darling and Blue Bird Mines were from the same ore lode, asserting that it was just as likely that miners in the Little Darling had encroached on Blue Bird ore as the reverse. Rossiter Raymond, for his part, worried that the faulting and shifting had made the situation far too complicated to explain succinctly and suggested to the Blue Bird owners that they build a defense based on an argument that the ore lode was continuous rather than faulted in any way. "I have seen nothing absolutely inconsistent with the existence of such a fault," Raymond wrote after reviewing his own notes and several affidavits. "On the other hand, I have seen and heard nothing inconsistent with the theory of a continuous vein, which if it prove[s] the true one will be much the simpler and more advantageous." Raymond noted that despite sustained attention to the problem, professional mining engineers could not arrive at a clear picture of the underground conditions; thus "a jury is almost certain to be muddled by it." After paying his experts and weighing the various possible outcomes, Van Zandt decided his best option was to avoid the courts altogether and make an offer to the owners of the Little Darling to buy out their claim, which he did in 1890.[50]

After absorbing mines and expanding their plants for years, several of the largest copper companies on Butte Hill—including the Anaconda, the Boston and Montana, and William Clark's Colusa-Parrot—had seemed to finally have organized their operations at a level appropriate to at least catch up with falling prices and diminishing ore quality, a scale of mining and smelting that appeared poised to stabilize somewhat by the mid-1890s. The complicated and tangled conditions of the ore lodes, however, presented new problems for company stability. Apex liti-

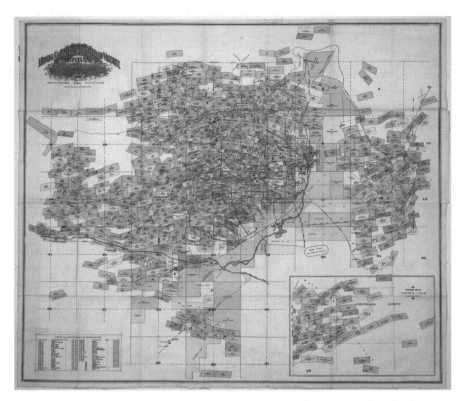

Map of mining claims for Butte City and vicinity, Montana, 1893. The tangle of surface claims, combined with the perplexing underground conditions that had resulted from the tumultuous geological history of the Boulder Batholith, made long-term underground ownership very uncertain on Butte Hill. Compiled and published by Baker & Harper; courtesy, Montana Historical Society Library Collection, map A-178, Helena.

gation and the time-consuming, costly, and fundamentally uncertain results it produced ensnared these companies in an uncertain struggle over ore ownership.

Mining companies filed numerous trespass lawsuits against other mining companies in the Butte Mining District during the 1890s, more often than not generating the same results as the Blue Bird versus Little Darling case had, with one company buying out the other instead of risking an unfavorable outcome in the courts. In 1896, however, the Boston and Montana Company filed the first of two large-scale trespass cases that would have long-term impacts on the industrial structure of the Montana copper industry. The case was filed in federal court against Augustus Heinze and his Montana Ore Purchasing Company. Heinze was a professionally trained mining engineer and former editorial staff member at the *Engineering and*

Mining Journal who had come to Butte in 1894 and worked for the Boston and Montana Company—the second-largest copper-producing company in the district at the time. The Boston and Montana Company had consolidated thirty mines on Butte Hill and since 1890 had sent a large share of its lower-grade ores to a large new ore-processing plant it had constructed on the shores of the Missouri River in Great Falls, Montana, northeast of Butte. Through his knowledge of the Boston and Montana Company's mines, his examination of the company's ore lodes, and a meticulous scouring of the district claim maps, Heinze had discovered that several odd-shaped pieces of property nestled among the large established claims had gone untitled.

When Heinze's father passed away, leaving him and his brother a significant fortune, he decided to invest it in mining, grabbing and locating several of the claims he knew were available. On one of them, the Rarus, Heinze began tunneling into the earth, eventually striking a thick vein of copper ore. The ore Heinze's miners struck and began extracting was located underneath the surface of Boston and Montana's Pennsylvania claim, adjacent to the Rarus. In 1896, Boston and Montana Company lawyers filed suit against Heinze for trespass.[51]

According to Heinze, his miners had followed an ore lode from its "apex" in his Rarus claim several hundred feet into the ground, where it ended up beneath the Pennsylvania claim; the vein in question belonged to him according to federal law. The court issued an injunction against Heinze's company, ordering him to stop mining until the case was resolved—which, in a move that seems to have sullied his credibility with later historians, he failed to do repeatedly. As the defendant in the case, Heinze was required to prove his innocence by establishing clear evidence that the ore lode whose outcropping appeared in his Rarus claim was the same ore lode his miners were extracting a few hundred feet under solid earth. This proved much more complicated than the federal law suggested it should be. To respond to the evidentiary requirements, Heinze had his engineers begin excavating what have become known as *litigation drifts*, mining tunnels whose only purpose was to establish the continuity of the ore lode from the surface outcropping to the place of mineral removal underground.[52]

After more than three years of study—during which time Heinze repeatedly violated the mining injunction against his company—the case went to trial. As in the earlier Blue Bird lawsuit, the maddeningly complex conditions in the highly fractured and fissured Boulder Batholith made for a confusion of facts. Heinze's engineers had constructed maps to depict the underground they had explored in their study, by which they made the argument that the ore lode under the Pennsylvania claim was in fact a continuation of the Rarus outcropping. Unfortunately for the clarity of their argument, however, the lode had fractured and shifted underground, so the ore lode found at depth was no longer physically continuous with the ore lode in the outcropping, but Heinze's expert witnesses argued that it was clearly the same original deposition of material. The prosecution's burden was merely to cast reasonable doubt on the facts presented by the defense, which it attempted to do by challenging the credentials of Heinze's experts, arguing over the meaning of mining terms and geological concepts used to describe the various underground structures and ore configurations, and presenting a panel of its own experts who used ore samples to make the argument that the ore being mined in the upper levels of the Pennsylvania claim was nearly identical in copper content to the ore removed by Heinze's miners from the lower level of the Rarus claim. After a year of conflicting testimony in the federal courts, in 1899 Judge Hiram Knowles remanded the matter back to the state courts without rendering a judgment because it involved two corporations that had both been incorporated in the state of Montana; as such, it was not an issue that could be decided in federal court.[53]

Not only were the ore lodes beneath Butte Hill Byzantine and confusing, the paths of these conflicts through the courts would be as well. Over the next seven years the Boston and Montana Company—which by the early twentieth century was another major company held by Amalgamated Trust—fought this battle in and out of state and federal courts, filing claims and counterclaims and even sending its miners into armed conflict in the enjoined but still working mines underground. Augustus Heinze, whose popularity among smaller mining operators and working miners became enormous in the years leading up to 1906, has been roundly criticized by mining historians for engaging in what some have called "courtroom mining," by which he disingenuously took advantage of the imprecise provisions of the 1872 mining law and the confounding underground conditions to exploit and extort from the largest producers on Butte Hill. However, given that Heinze was sued first by the Boston and Montana Company and subsequently spent tens of thousands of dollars constructing a legal defense for a case that was never resolved as a legal issue, the harshest and least generous conclusions reached by historians are impossible to confirm. What is clear is that such impossibly complicated battles over underground ore lodes created a dangerous new uncertainty for the large-scale mining companies trying to maintain a footing in the volatile copper market.[54]

The year after the Boston and Montana Company filed suit against Heinze's company, William Andrews Clark's Colusa-Parrot Mining Company filed a similar, although more complicated, trespass case against the Anaconda Copper Company. Colusa-Parrot attorneys claimed the Anaconda Copper Company was mining in their property, more than 1,500 feet underground, which belonged to them under the provisions of the 1872 mining law. They agreed that Anaconda miners had followed an outcropping from the surface of the Anaconda claim into the ground beneath the

Parrot claim, but they asserted that this outcropping was merely a branch of a larger single ore lode that united beneath the Colusa-Parrot claim. In such circumstances, the 1872 mining law gave ownership of the entire vein beneath the point of division to the prior owner between the two surface claims. Since William Clark had located his claim back in 1872, before heading to the School of Mines at Columbia College, and Michael Hickey had located the Anaconda claim in 1875, according to the provisions of the mining law, Anaconda's "apex rights" did not apply on the lower section of this ore lode. This was a significant charge, whose outcome could potentially change the shape of the Butte copper industry. In effect, Clark's company was making claims on the district's largest copper producer's largest and most productive ore lode, a claim the Anaconda Copper Company took very seriously and with every intention of successfully challenging.[55]

Sparing no expense, after entering into a consent decree with Colusa-Parrot to limit the employment of experts in an effort to keep the trial's length and costs within reason, Anaconda proceeded to spend more than $250,000 to hire three of the era's most prominent mining experts: Nathan Shaler, a professor of geology at Harvard University; Clarence King, founder and former head of the US Geological Survey; and Rossiter Raymond, founder and former president of the American Institute of Mining Engineers (among his many accomplishments). Building from a defense first developed in the Blue Bird case, the men constructed elaborate maps and a three-dimensional scaled model of the underground vein system as they envisioned it to prove that the Anaconda ore lode did not—indeed, could not—intersect with the Colusa-Parrot ore lode. Their contention was as simple as the original complaint was complicated. The men claimed to have found—and they depicted it visually—an anomalous "blue vein" (colored bright blue in both the drawings and the model) they said bisected the Colusa-Parrot lode somewhere above the location of the purported juncture, thus rendering the meeting of the two veins a physical impossibility. On the witness stand, King claimed the blue vein could be identified by its unique mineral characteristics. Shaler then identified a secondary fault zone running across the entire mining district, in the fissure of which the blue vein formed, cleanly separating the Anaconda and Colusa veins. Raymond testified that the blue vein existed exactly as the full-color model depicted it, a fact he had determined after sixty days of examining the topography and the Butte mines.[56]

The Colusa-Parrot attorneys did their best to cast doubt on the "blue vein" theory. When, for example, they pressed Nathan Shaler to identify the boundaries of the so-called blue vein, Shaler balked. "It is like the case of looking at a cloud and telling just where the cloud is and the sky is," he answered obliquely. When Clarence King was pressed to defend his claim that the blue vein had a different mineral content than the other ore lodes, he had to admit that the distinguishing characteristics (specifically, its mineral content) were the same as those of the Anaconda and Colusa-Parrot veins. "Q. Then you find no metal in what you call the blue vein that is not found in some quantity in and in some place in what you call the Anaconda and the Colusa-Parrot? A. Quite right." King followed up by saying that the real difference could be found in the quartz content of the blue vein, but again, when pressed, he had to admit that such variations in quartz content existed within both of the other veins.[57]

After a lengthy trial that extended through 1899 and into 1900, Judge Knowles—the same judge who would remand the Boston and Montana Company case back to the state court a year later—rendered his decision in favor of Anaconda. Knowles admitted to his difficulty in coming to a conclusion and to the generally unscientific approach he took in reaching it: "While the evidence of the witnesses for the plaintiff would seem to deny the existence of this Blue Vein, it does not commend itself to me as being so clear, direct and positive as that of the defendant."[58] In fact, despite the legal necessity for Anaconda to prove its ownership, the company seems to have succeeded as much because it could present an argument that was easier for the judge to understand and see than the one presented by Colusa-Parrot's experts. The Anaconda legal team did the company's long-term prospects a great service by including in their successful defense "that the Anaconda vein extends throughout the length of the Anaconda claim from end to end; that in its dip it passes out of the Anaconda through the Never Sweat [sic] and into the claims of the plaintiff and even through them and beyond."[59] Marcus Daly had passed away in New York City in 1900 before the case had been decided, but never again would the company he built have to worry about access to its ore lode underneath Butte Hill.

As in all cases that riddle the courts and put ownership of underground reserves at risk, the Anaconda case was not decided on a clear rendering of unambiguous or even visible facts but instead rested on hypotheses, speculation, and projection couched in the guise of scientific knowledge. Companies hoped they could bring to the stand the most convincing story with the most apparently trustworthy experts; they did not, in the end, seek accurate science, merely better-illustrated stories about the evidence. In the first comprehensive history of American mining, Thomas Rickard reflected on the trials in Butte with no small amount of disdain: "Lawyers celebrated for their forensic skill and geologists honored for their scientific knowledge were to be seen and heard as they played their parts [and] posed as consultants to Wise Providence in the creation of the ore deposits." Rickard was a member of the American Institute of Mining Engineers (AIME) who began his career during the final years of Rossiter Raymond's life, and his *A History of American Mining* was published first for AIME by the Scientific Publishing Company before being reprinted by Macmillan, so it is fair to assume that he was generously predisposed toward the ore trespass lawsuits at the turn of the century in Butte. Yet he ultimately concluded that for all the theories and drawings and models, nothing of scientific value was produced by any of the witnesses for any of the companies. "It is a significant fact," Rickard concluded, "that the interesting evidence given by the geologists so positively in the courts is not quoted in any geologic treatise or in any scientific discussion, for the simple reason it is suspect."[60]

Where the gold rush activities had produced a geography of minerals that was beyond the gold miners' abilities to fully utilize and the lode mining industry had similarly launched a pattern of iterative expansive growth that seemed always to put mineral profits just beyond the reach of western mining investors, copper exploitation, in contrast, came at the hands of a business strategy that seemed to promise the transcendence of these structural difficulties. The financiers who developed copper properties in the 1890s had gained control of a European technology through the commodification of the artifacts and the knowledge of advanced mining. They then sought to solve the problem of iterative growth that had plagued the silver producers by expanding their mass-production capacity and absorbing other related businesses and industries necessary for their ore production processes under a single management. By consolidating all the mines, the financiers intended to avoid any conflicts over ore ownership; by consolidating ore production, they hoped to manage conflicts with other users of the common land, air, and water resources. They did not anticipate the drawn-out conflicts over ore ownership that would continue to challenge some of the largest companies until 1906, but the Anaconda victory against Calusa-Parrot in 1900 resolved that uncertainty for Amalgamated Trust—which owned the Anaconda Mine and mills—and cleared the way for an unobstructed consolidation of the district by Amalgamated.

PATTERNS OF MODERN BUSINESS

Curiously, the success of Amalgamated Trust stemmed largely from the industrial system Daly and Haggin had built in western Montana. While the flood of copper on the market in the 1890s must have seemed like an abundant natural resource, all of that copper at such low prices was anything but a natural fact, as we have seen. Rather, it had been produced. By the mid-1890s copper was produced everywhere at the direction of an elaborate coordination of resources by a management hierarchy, a business strategy designed to contain the costs that had challenged silver lode miners in the late 1870s.[61]

Daly and Haggin had led the way in Butte with their diverse investments in related businesses and industries. As Sargeant noted in his report to James Hague on company operations in the 1880s and 1890s, this diversity was designed to reduce costs by taking control of the supply stream, eliminating delays, and removing middlemen who took their own share of the action. Other copper producers had taken similar steps: integrating mining claims, investing in forest resources, and investing in new large ore-processing plants and town site lots to gain political and economic control over the landscapes produced by their processing activities.[62]

The system building and economic integration performed by Daly and Haggin and other large producers in the Butte copper industry follow the pattern of the management hierarchies and modern business strategies described in the work of Alfred Chandler. In his 1977 study of business formation in the United States, *The Visible Hand*, Chandler described management hierarchies emerging in business institutions as the rational solution to the coordination of movement across the US geography on the transcontinental railroads. Attention to units of management and a hierarchy of communication among them allowed railroad companies to schedule and coordinate freight going to and from multiple locations across multiple time zones. The "management hierarchy" innovation, Chandler argued, was then replicated in a wide range of mass-production industries, from steel manufacture to food processing, giving birth to the modern corporation. Management hierarchies internalized the movement of goods, replacing the invisible hand of the market with the rational mind of the manager and creating enormous cost savings for businesses in the process.[63]

Chandler attributed these changes to factors external to the business enterprises themselves. First, he identified growing market demand as a major pull factor on institutional growth—Americans wanted to buy more stuff at a lower cost; business growth was simply a response to that demand. The increasing scale of production had the secondary effect of driving down prices, which further increased demand. Second, Chandler identified innovations in communication, transportation, and production technologies as a major push factor—being able to stay ahead of the market by having distant information instantly and being able move large volumes both quickly through the production process and over long distances at affordable rates also encouraged firms to expand. Throughout the 1880s and 1890s and peaking in the early twentieth century, large economic institutions came to dominate the US economy as they responded to these factors by expanding their reach.

According to Chandler, the expansion came in two forms, which he called *horizontal* and *vertical* integration. Some firms expanded horizontally by purchasing or driving out of business competitors doing the same thing. The textbook example was undertaken by Andrew Carnegie during the late 1880s to build an empire of steelmaking by taking control of most of the US steel production capacity and then using this bottleneck in the industry to exercise control over iron mining and steel fabrication and distribution. Other firms expanded vertically by gaining control over supply lines at one end and distribution networks at the other. Chandler points to several large food-manufacturing firms that undertook this "vertical integration" first. According to Chandler, the most successful corporations in the twentieth century had grown out of a combination of these strategies by the end of the nineteenth century and usually came to enlist both.[64]

The similarities between Chandler's elegant and highly influential thesis about the evolution of business enterprises in the United States and the pattern of business growth in Montana's copper industry are curious. Chandler not only neglected to analyze the patterns at play in Montana in the 1880s and early 1890s—the same years his iconic industries were making their most important changes—but he specifically stated that the mineral production industry and in particular the Anaconda Copper Company had arrived late to the patterns of the modern business enterprise. The industry failed to do anything of meaningful institutional significance until the World War I era, when it vertically integrated forward into consumer fabrication and mass marketing.[65]

As this chapter has detailed, however, the companies in the Butte copper industry that survived after 1890 had undertaken significant horizontal and vertical integration of mines and supplies as they worked to keep costs low enough to profit from their production systems. The largest companies absorbed dozens of smaller copper and silver mines on the hill as they grew. In the wake of the mine trespass cases, Michael Malone has written that the integration of all Butte mines into Amalgamated Trust—eventually renamed the Anaconda Copper Company for its flagship mining property and ore-processing plant in Anaconda—was "inevitable. Scattered ownership guaranteed only waste, contention, and endless litigation." The total horizontal integration of the Butte mines, which began in 1899 and was completed by 1910, had been made economically necessary because the ores were unpredictable in shape as well as quality and because aggressive competition added excessive costs to doing business in the region. The successful companies had first integrated vertically—owning timber and mills, fabrication and machine shops, coal mines and railroads, wholesale importers, and a vast gaggle of land and service industries. They had then integrated horizontally on Butte Hill, consolidating mine ownership through lawsuits and purchases and, eventually, smelting capacity. The costly lawsuits over the future ore reserve inspired and encouraged the integration of the entire copper region into Amalgamated Trust, whose managers knew the only way to stabilize the price of copper was to control its production in Butte, Anaconda, and Great Falls.[66]

In fact, there was no shortage of horizontal and vertical integration in the Montana copper industry during the key period of institutional change delineated by Chandler's work, but there were causal differences. Modern business and managerial hierarchies developed in Montana in reaction to the geologic uncertainties of lode mining: they sought the means to control transaction costs because they could not control raw material costs, and they could not control raw material costs because of the nature of ore and the ways ore had become part of the US political economy. Integration did not stem from significant increases in market demand, although it was certainly facilitated by improvements in communication and transportation. Copper producers had flooded markets in the 1880s and 1890s as a competitive strategy, an effort to price others out of business. Falling prices encouraged companies to build larger processing plants for economies of scale and to keep up with the diminishing quality of ore coming out of the ground. The few large firms that survived the 1890s depression had taken control of timber reserves, copper mines, processing works, and railroads. Only local vertical integration afforded the means to mine copper profitably. By the turn of the century, even that was not enough, and the entire region would be centralized under the control of a single management.[67]

In Montana, this meant not only the rapid consolidation of the Butte mines into the hands of Amalgamated Trust in the early twentieth century but also a consolidation of processing and smelting from a variety of large mills in Butte and on the northern hills of Anaconda into a single, massive, sprawling ore production facility—the Amalgamated (Anaconda) Trust's Washoe Reduction Works, a 250-acre ore-processing plant with the staggering ability to process more than 6,000 tons of raw ore a day.

By the dawn of the twentieth century the institutional dynamics of the American copper industry, which had followed the suggestions of professional mining engineers about planning and measuring and testing, had not only outgrown and then fully absorbed the interests of these engineers, it had also outgrown and exceeded the capacity of even the wealthiest and most experienced of the first-generation western mining capitalists. It had become not only a national but a world industry whose output affected markets in New York, London, and Tokyo and whose material product made possible a new age, a revolution in power and transformation of modern life over the course of the twentieth century.

But it also became a landscape-scale metabolism. The new centralized facility at the Washoe Reduction Works would put a greater, more concentrated volume of harmful material into the waters and air of the Deer Lodge Valley than any region of the world had ever experienced. The logical culmination of the mining practices that had taken root in the US West encompassed massive mass-production ore-processing facilities, harnessing the power of coal and converting medium- and low-grade minerals into commodity metals. These were the processing facilities of mining lore, running day and night, producing the sights and sounds and smells that marked the mechanical, industrial, superhuman qualities of the modern smelter. When the

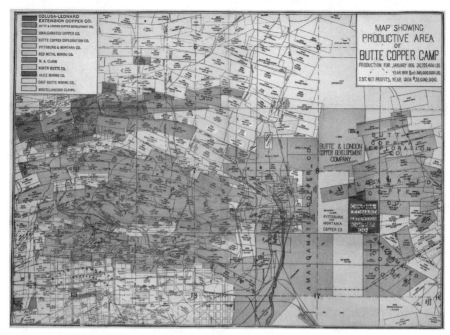

Map showing productive area of the Butte copper camp. By 1906, all Butte mines had been consolidated by Amalgamated Trust. Among other accomplishments, Amalgamated Trust (soon renamed the Copper Company) succeeded in gaining spatial control over the entire group of valuable and potentially valuable claims on Butte Hill. Once and for all, access to the underground ore lodes was assured. Courtesy, Montana Historical Society Library Collection, map C-494, Helena.

Washoe Reduction Works was completed in 1903, it was the largest plant of its kind in the history of the world.

The Washoe plant represented the logical culmination of the uncertainties of mineral production, the low price of copper, and the generally extensive deposit of copper in the Boulder Batholith beneath Butte. But the plant was not located in unused country. It sat at the southern reaches of the Deer Lodge Valley. The grassy plain nestled between ranges of mountains had been agricultural land ever since John Grant and the Stewart brothers had first run horses and cattle in the 1850s. The land on which James Fisk described fattened cattle and verdant crops in the 1860s was still predominantly used for cattle grazing in the early twentieth century, but it had included row crops and milk cows since the 1870s. Mining had been part of the Deer Lodge Valley economy since the first rush to Gold Creek in 1862, and no one questioned its presence. But the farmers and ranchers who continued to work the valley lands had never before been subject to the volume, concentration, or character of effluents that poured out of the Washoe plant and into the ecosystems on which their livelihoods depended. The farmers' reaction would constitute a final and critical uncertainty in which the legitimacy of competing landscape visions would come before the courts.

NOTES

1. F. E. Sargeant, *F. E. Sargeant's Report of the Anaconda Copper Mining Company,* Mining Reports of Individual Mines, Companies, and Properties, 5, Anaconda Mines, Butte, Montana, Reports, Statements, Correspondence, etc., 1882–1906, Box 14, James D. Hague Collection, HHMC.

2. David Alt and Donald W. Hyndman, *Roadside Geology of Montana* (Missoula, MT: Mountain Press, 1986), 3–14, 133–35, 140–43.

3. Of course, none of this could be known prior to the mining that took place in Butte, and much of the knowledge was derived during trespass cases when companies were forced to prove their ownership of underground ore deposits and thus conceptualize them for the court for the first time.

4. E. G. Brown, "The Ore Deposits of Butte City," in *Transactions of the American Institute of Mining Engineers* (New York: Scientific Publishing, 1894), 543–58.

5. Davis, *Story of Copper.*

6. *Transactions of the American Institute of Mining Engineers*, vol. 1, May 1871–February 1873, 320; "Review of the Metal Market for 1878," *Engineering and Mining Journal* 29 (1879): 38–39.

7. "Montana: Butte District," *Engineering and Mining Journal* 32 (1882): 112–13; Malone, *Battle for Butte,* 25–27.

8. "Anaconda Property in Account with J. B. Haggin," P. Mining Reports of Individual Mines, Companies, and Properties, 5, Anaconda Mines, Butte, Montana, Reports, Statements, Correspondence, etc., 1882–1906, Box 14, James D. Hague Collection, HHMC; "Mining Notes: Montana," *Engineering and Mining Journal* 36 (1883): 388; Malone, *Battle for Butte,* 24–25.

9. Malone, *Battle for Butte,* 27; Glasscock, *War of the Copper Kings,* 52–53.

10. James Hague, "Ledger Account J. B. Haggin," P. Mining Reports of Individual Mines, Companies, and Properties, 5, Anaconda Mines, Butte, Montana, Reports, Statements, Correspondence, etc., 1882–1906, Box 14, James D. Hague Collection, HHMC.

11. "Mining Notes: Montana," *Engineering and Mining Journal* 32 (1882): 113; Malone, *Battle for Butte,* 27–28.

12. Malone, *Battle for Butte,* 29.

13. Hague, "Ledger Account J. B. Haggin," HHMC.

14. Malone, *Battle for Butte*, 29; James J. Hague, *Anaconda Mining and Smelting Properties*, submitted to James B. Haggin, December 24, 1889, 6, P. Mining Reports of Individual Mines, Companies, and Properties, Box 14, James D. Hague Collection, HHMC; "Anaconda Property in Account with J. B. Haggin," HHMC; Edward Dyer Peters, "Mines and Reduction Works," Reports, Statements, Correspondence, etc., 1882–1906, Box 14, James D. Hague Collection, HHMC.

15. "Montana," in Browne, *Report*; Eilers, "Western Montana," 168; James Fell, *Ores to Metals: The Rocky Mountain Smelting Industry* (Lincoln: University of Nebraska Press, 1979), 140–41.

16. "Mining Notes: Montana," *Engineering and Mining Journal* 28 (1879): 339: "Mr. C. T. Meader, of Butte, has commenced the erection of copper smelting works for the handling of argentiferous copper ores. The company Mr. Meader represents will complete the erection of the works this fall. The works will be located about a mile and a half northeast of town, near the Colusa and other lodes owned by the company. The company has entered the ore market of Butte as a large purchaser, having recently acquired the entire dump of the Gagnon mine for a consideration of $15,000." See also "Mining Notes: Montana," *Engineering and Mining Journal* 29 (1880): 134, and 27 (1879): 452; Malone, *Battle for Butte*, 21–22; Quivik, "Smoke and Tailings," 110–18; James Douglas, "Letters from the West: IX," *Engineering and Mining Journal* 33 (1882): 219.

17. Edward Dyer Peters, *Modern Copper Smelting*, 7th ed. (New York: Scientific Publishing, 1895).

18. Michael Tanzer, *The Race for Resources: Continuing Struggles over Mineral and Fuels* (New York: Monthly Review Press, 1980), 42, 57; Quivik, "Smoke and Tailings," 155–64.

19. "The Deer Lodge Smelters," *New North-West* [Deer Lodge, MT], May 18, 1883, 5.

20. Ibid.

21. Malone, *Battle for Butte*, 35–37. Quivik, "Smoke and Tailings," 157, referred to the decision to build such a large copper smelter as "one of the most unfathomable episodes in early Butte history" because the men who invested had only developed precious metals, the domestic market in copper was more than adequately served by the Michigan mines, and the expense of a plant to treat the complex ores seemed to have suggested conservative development. But given how much ore Daly knew he had and knew he had to sell, producing as much as possible all at once was in fact the economically rational choice.

22. H. Minar Shoebotham, *Anaconda: Life of Marcus Daly the Copper King* (Harrisburg, PA: Stackpole, 1956), 88–97; Malone, *Battle for Butte*, 29–31; "The Deer Lodge Smelters," *New North-West*, May 18, 1883, 5.

23. "General Mining Notes: Montana," *Engineering and Mining Journal* 36 (1884): 132, 200; "Anaconda Gold and Silver Mining Company, Anaconda Smelting Work. Statement of Cost and Operating Expenses," Box 446:5, Manuscript Collection 169, MHS; Peters, "Mines of Butte," 384; Titus Ulke, "Characteristic American Metal Mines: The Anaconda

Copper Mines and Works," *Engineering Magazine* (July 1897): 516; James Douglas, *Cantor Lectures on Recent American Methods and Appliances Employed in the Metallurgy of Copper, Lead, Gold, and Silver* (London: William Trounce, 1895).

24. Henry B.C. Nitze, "A Description of the South Welsh Method of Copper Smelting as Practiced by the Baltimore Smelting and Rolling Co., Baltimore, Md.," BS thesis, Lehigh University, Bethlehem, PA, 1887.

25. "The Anaconda Mines, the Greatest Copper Mine and the Largest Smelters in the World," *Financial Times,* October 23, 1891; James Hague, "Exhibit 'C'—Anaconda Copper Mining Company's Works," 5, Chicago, P. Mining Reports of Individual Mines, Companies, and Properties, Anaconda Mines, Butte, Montana, Reports, Statements, Correspondence, etc., 1882–1906, Box 14, James D. Hague Collection, HHMC.

26. "F. E. Sargeant's Report of the Anaconda Copper Mining Company," 13, P. Mining Reports of Individual Mines, Companies, and Properties, Anaconda Mines, Butte, Montana, Reports, Statements, Correspondence, etc., 1882–1906, Box 14, James D. Hague Collection, HHMC ; "General Mining Notes: Montana," *Engineering and Mining Journal* 50 (1890): 15; "Anaconda, St. Lawrence & Never Sweat—Tons of Ore Shipped to Anaconda," single sheet in sleeve of notebook, "R2—Anaconda Mines, September 18–21 [1895]", R. Notebook—Mining, General Notes, James D. Hague Collection, HHMC.

27. For a general description of these holdings, see "F. E. Sargeant's Report," 13–19.

28. Quivik, "Smoke and Tailings," 54 (quote), 128–77.

29. Joseph Newton, *Metallurgy of Copper* (New York: John Wiley and Sons, 1942), 481; "The Low Price of Copper," *New York Times,* December 18, 1885.

30. Hague, *Anaconda Mining and Smelting Properties*, HHMC.

31. Arguably, the automobile, atomic bomb, and personal computer have an equally symbolic status. Paul Israel, *Edison: A Life of Invention* (New York: John Wiley and Sons, 2000); Hughes, *Networks of Power*; and Langdon Winner, *The Whale and the Reactor: A Search for Limits in an Age of High Technology* (Chicago: University of Chicago Press, 1986) address this theme in different ways.

32. Malone, *Battle for Butte,* 35.

33. William Cronon, "Kennecott Journey," in William Cronon, George Miles, and Jay Gitlin, eds., *Under an Open Sky: Rethinking America's Western Past* (New York: W. W. Norton, 1992), 40.

34. Hughes, *Networks of Power*, 39.

35. Charles K. Hyde, *Copper for America: The United States Copper Industry from Colonial Times to the 1990s* (Tucson: University of Arizona Press, 1998), 49.

36. William Robbins, *Colony and Empire: The Capitalist Transformation of the American West* (Lawrence: University Press of Kansas, 1994), esp. chapter 5, "The Industrial West: The Paradox of the Machine in the Garden."

37. Hughes, *Networks of Power.*

38. Horace J. Stevens and Walter Harvey Weed, *The Copper Handbook: A Manual of the Copper Industry of the World*, vol. 9 for the Year 1909 (Chicago: M. A. Donohue, 1910), 1605.

39. Malone, *Battle for Butte*, 36–38.

40. Hughes, *Networks of Power*, 29.

41. Israel, *Edison*, 184.

42. See Menlo Park Notebook no. 1 (November 28, 1878–July 24, 1879), section on wire calculations, in Hughes, *Networks of Power*, 29.

43. Hughes, *Networks of Power*, 29–42.

44. Malone, *Battle for Butte*, 38–39.

45. Hague, *Anaconda Mining and Smelting Properties*, HHMC.

46. Anaconda Mining Company, Anaconda, Montana, Financial Report, 1891 to 1894, 5, P. Mining Reports of Individual Mines, Companies, and Properties, Anaconda Mines, Butte, Montana, Reports, Statements, Correspondence, etc., 1882–1906, Box 14, James D. Hague Collection, HHMC; Malone, *Battle for Butte*, 131–39.

47. Rickard, *A History*, 362.

48. "Trial Transcript," Colusa-Parrot v. Anaconda, 188–89, March 21, 1900. Many of these geological facts came to light for the first time in the court cases. See, for instance, "Trial Transcript," Boston and Montana Copper & Silver Mining Company v. Montana Ore Purchasing Company, Civil Case File #35 (Boston and Montana v. Montana Ore, #34), and *Transcript of Testimony*, Colusa-Parrot Mining & Smelting Company v. Anaconda Copper Mining Company, NARA-PR. See also Lester G. Zeihen, Richard B. Berg, and Henry G. McClernan, "Geology of the Butte District, Montana," in Donald L. Biggs, ed., *Geological Society of America's Centennial Field Guide* (Boulder: Geological Society of America, 1987), 57–61.

49. D. W. Brunton, *Technical Reminiscences* (New York: Mining and Scientific Press, 1915), 12.

50. Ibid., 12; Affidavit of James D. Hague, and Letter from Rossiter W. Raymond to Robert E. Booraem, November 30, 1889, both in 17, P. Mining Reports of Individual Mines, Companies, and Properties, Blue Bird Mine—Butte, Montana, Report, Correspondence, Affidavits, Maps, 1888–91, Box 15, James D. Hague Collection, HHMC.

51. "Complaint," Boston and Montana Consolidated Copper & Silver Mining Company v. Montana Ore Purchasing Company, Civil Case File #34, 1890–1912, US District Court, Butte, Record Group 21, NARA-PR. On Heinze, see Malone, *Battle for Butte*, 49–53.

52. "Answer," Boston and Montana Consolidated Copper & Silver Mining Company v. Montana Ore Purchasing Company, Civil Case File #34, 1890–1912, US District Court, Butte, Record Group 21, NARA-PR; Malone, *Battle for Butte*, 168–73.

53. "Trial Transcript," Boston and Montana Consolidated Copper & Silver Mining Company v. Montana Ore Purchasing Company, NARA-PR.

54. Malone, *Battle for Butte*, 159–89.

55. Ibid., 144–48.

56. *Transcript of Testimony*, 800–1023, Colusa-Parrot Mining & Smelting Company v. Anaconda Copper Mining Company, NARA-PR.

57. Ibid., 589–1394, 638 (Shaler), 393, 397–99 (King).

58. "Opinion of the Court," 7, Colusa-Parrot v. Anaconda, filed June 2, 1900, Civil Case File #61, 1890–1912, US District Court, Butte, Record Group 21, NARA-PR.

59. "Answer by the Defendant," 4, Colusa-Parrot v. Anaconda, Civil Case File #61, 1890–1912, US District Court, Butte, Record Group 21, NARA-PR.

60. Ibid., 364; Rickard, *A History*, 362.

61. Winner, *Whale and the Reactor*.

62. "F. E. Sargeant's Report," HHMC.

63. Alfred Chandler, *The Visible Hand: The Managerial Revolution in American Business* (Cambridge, MA: Harvard University Press, 1977).

64. Ibid.

65. Ibid.

66. Malone, *Battle for Butte*, 131.

67. LeCain, *Mass Destruction*.

The great bulk of the metallic ores, when in their natural
situations, constitute a most heterogeneous mixture, in
which the really valuable metal exists only in a small
proportion, chemically combined with one or more
mineralized substances, and entirely intermixed with
sparry and earthy matter, and ores of inferior metal.

John R. Leifchild, *Cornwall: Its Mines and Miners* (1855)

Between Butte and Missoula the lovely and limpid river
which Lewis and Clark marveled at is now transformed
into a muddy, slimy, dirty river, as foul looking as the
water of the Chicago, the Tiber or the Yarra-Yarra.
Along its banks have grown up communities rich, and
even luxurious, and the wondrous prosperity of Butte is
responsible for the destruction of the beauty of the once
lovely Hell Gate River.

Pete O'Farrell, *Butte: Its Copper Mines and Copper Kings* (1899)

The amount of arsenic present is very large.

Horace J. Stevens and Walter Harvey Reed, *The Copper Handbook* (1910)

BY THE TURN OF THE twentieth century, the mass production of copper in oversized ore-processing plants comprised the material heart of a new energy regime in the
United States. In urban centers back East, in Europe, and
in Japan, electrification was changing the face of domestic life, work life, and leisure time. Alongside the internal combustion engine, electrification systems created
the technological context for vast changes in US culture,
contributing to a growing emphasis on the domestic and
institutional consumer—predominantly urban and aspi-
rationally middle class. The rural mass production of metals—copper, zinc, lead, iron, silver, and eventually bauxite and uranium—represented the necessary material
precondition and continued prerequisite for the generation of this mode of production. It was an energy regime
of intricate machines made of mineral-based materials. It
was a mining society.[1]

FOUR

*The Ecology of
Ore Processing*

*Pollution and the Law in
the Deer Lodge Valley*

DOI: 10.5876/9781607322351.c04

Electricity was first used in the urban streetcar systems constructed in the 1890s but was then extended into the apartments and houses of city dwellers served by electrical-generation utility companies, which built vast networks of municipal and domestic power throughout the developed world (indeed, defining the developed world as such) during the first quarter of the twentieth century. This development produced growth seemingly without end. The presence of electrical power as a regular part of domestic life encouraged markets in new domestic electrical goods. Middle-class consumption became an increasingly important and distinct source of economic growth. Electrical systems and many of the technologies they supported became so much a part of everyday existence that modern life became virtually unthinkable without them. This world-historic technical accomplishment was not only made possible by the profligate production of copper; it also, in turn, linked the western mountain copper production centers ever more intimately with the environment of modern urban places and assured a regular market in copper for years to come. By the turn of the twentieth century, the volume of trade in copper products attracted new levels of finance in the form of trust companies eager for a chance to participate in the apparent riches of the western metal production that supported the new technologies of the urban environment. For eastern financiers with extensive fortunes and, more important, access to the fortunes of others, copper mining and processing became a gamble of significant consequence.

Despite access to vast sums of liquid capital and the best mining and engineering minds on the planet, the new investors still had to produce copper at a low enough cost to return a regular profit. This persistent challenge had already placed Montana's large-scale copper industry on a path of endless and costly reinvestments. Ore qualities and market prices could not be trusted. During the entire period of initial consolidation in the 1880s and 1890s, the large producers in Butte had battled over underground ownership of ore lodes, and Butte residents had objected to the pall of smoke polluting their valley air. One of the many obstacles to profitability had been the limits on the space and volume of throughput imposed by the size of the Summit Valley. By the beginning of the twentieth century, centralized processing appeared to be the only solution to the ongoing market chaos. Copper's new role in urban life had inspired confidence in a steady market for the metal, and finance-driven consolidation began in Butte at the hands of Amalgamated Trust.

The key to the Amalgamated production strategy had been the vast landholdings it had acquired when it purchased the Anaconda Copper Company from Marcus Daly and James Haggin in late 1899. Indeed, the trust had built the Washoe Smelter on the 250 acres of unused company lands downstream and uphill to the west of the town of Anaconda, at the farthest southern reaches of the Deer Lodge Valley. Key to cost reduction, the Washoe Reduction Works was built to process 6,000 tons of low-grade copper ore a day, a scale of ore processing unprecedented in human history. Starting in 1900, most of the ore produced out of Butte Hill would be processed through this single ore-processing plant. The consolidation of mine output solved the problem posed by Butte's overcrowded air and water, and it offered a scale that would necessarily inflate marginal savings and generate profitability on a low-cost metal. In other words, Amalgamated managers believed they could achieve the long-sought-after profitability in low-grade copper ore production by building a gargantuan mass-production ore-processing and copper-refining plant in the southern Deer Lodge Valley.

Mass production of ore gained favor among investors like the Amalgamated men because it achieved an economy of scale; it saved money on transaction costs by moving things in bulk, and small savings multiplied by the thousands of dollars. Entire trains of ore could be dumped into the hopper at the top of the concentrator building, saving on both fuel and labor costs. Trainloads of calcines could be automatically dumped into smelting ovens and other furnaces throughout the sprawling campus of industry that comprised the Washoe Reduction Works. The production processes also saved money because they concentrated wastes, presumably making them easier to manage and control. The facility adjoined adequate disposal space for one very large slag pile, another large piece of land on which the company had built berms for huge tailings ponds. In addition, it had endless mountain breezes to carry away the smoke from the roasting and smelting ovens, diffusing it into the atmosphere. Smelters had faced a challenge from Butte residents in the 1880s and 1890s because of copper smelting. Washoe had been designed to eliminate these challenges. In many ways, it would be fair to say that Amalgamated Trust had finally mustered enough resources to solve the problem of the ores. Piece by piece, mine by mine, the trust brought known, well-developed mining properties under a unified management. Consolidating all of the mines into a single processing plant cut costs even further. Everything looked good on paper.

The pure economics of Amalgamated Trust's copper production system—the consolidation of Butte's entire body of ore lodes, the centralization of ore processing and mineral smelting on a scale beyond imagination, and the marketing of huge quantities of raw copper product (and, soon, fabricated copper product)—indicated profitability. But the laws of physics caught up with the men of finance once the Washoe facility went into production. While little was new in the milling and smelting techniques installed in the Washoe Reduction Works since the various processes and technologies had been used for more than twenty years in the region and longer in other places, the scale was unprecedented. The savings afforded by large-scale mass ore processing required the ongoing ability to dispose of almost 95 percent of the rock material brought to the processing facility in the first place. All but 5 percent of the original material dumped in the concentrator left the plant as waste—sluiced to a tailings dump, poured out as slag, or lofted in searing hot gases up one of the chimneys—and this waste was hazardous and toxic. As the annual production of copper increased in step with the spread of electrification, the production of waste grew in kind. Amalgamated's managers suddenly found themselves generating more waste from a single facility than had ever been generated in one place all at once—so much waste, in fact, that the entire landscape in the Deer Lodge Valley began to change. These changes, expressions of an unhealthy ecosystem, precipitated new uncertainties for the metal producers to confront.

BUTTE COPPER ORES

Calling the mineral deposits that crystallized into the fissures beneath Butte Hill "copper" ores is a bit of a misnomer. While some of the original material saved by William Clark and Marcus Daly and shipped east for processing in the 1880s contained mostly copper—the copper glance, or chalcocite as it has been named by geologists, contained on average 67 percent copper by volume and sometimes as much as 80 percent by weight—those minerals did not comprise the entire mass of the ore lode or even a significant portion of the ore mined in the Butte copper zones. The geological deposition of "copper" minerals into fissures does not create a uniform ore material, as the superheated solution is not itself uniform. In Butte and other, similar copper ore deposits around the West, the largest share of material in the ore was in fact silica and sulfur. The chemistry of ore deposition caused new chemical combinations through precipitation, resulting in sulfur compounds in the valuable portion of the ore material—hence the term *sulfide* deposits.[2]

Despite the variety and uncertainty, however, the ore lodes contained pockets and chutes and seams of minerals commonly known as "copper" ore—bornite, chalcopyrite, enargite, and tennantite were the main sulfide minerals exploited from these enriched areas of the ore lodes—but again, "copper" was not quite accurate since even the richest material going into any of the ore-processing plants in the early years contained just under 30 percent copper, and by the early twentieth century the richest ores barely exceeded 15 percent copper. Moreover, these first-class ores, as they were called, comprised no more than 10 percent of the ore mined out of the Butte mineral zone. The remaining 90 percent of the ore contained the much lower average of 5 percent copper. These second-class ores, in other words, were mostly not copper. The ores were, in fact, mostly granite and silicate, which comprised about half of all material excavated and sent for processing.

The remaining material that did not include copper had a generally regular ratio of several common elements. The most abundant material in the remaining ore was

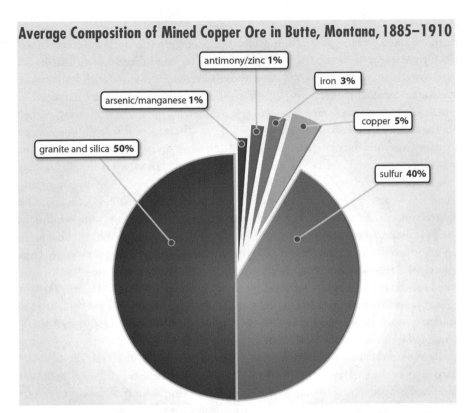

Average Composition of Mined Copper Ore in Butte, Montana, 1885–1910

antimony/zinc **1%**

iron **3%**

arsenic/manganese **1%**

copper **5%**

granite and silica **50%**

sulfur **40%**

On average, between 1880 and 1910 Butte's copper ores contained mostly substances that were not copper, many of which oxidized and volatized into the surrounding landscape, with dangerous impacts (source in note 3).

sulfur, a non-metal element that is not only abundant on earth—about five time more abundant in the lithosphere than copper—but is the tenth-most-abundant substance in the universe. Sulfur made up as much as 40 percent of the total volume of material mined from the ore lodes. Sulfur also formed the base and the largest share of each of the major "copper" minerals found latticed throughout Butte's ore lodes, binding chemically with the other materials (iron, copper, manganese, arsenic, antimony, zinc) and forming distinctive crystalline mineral deposits where enough material was present to precipitate into regular compounds. On average, iron was the next-most-abundant material in the remaining ore lode. Iron wasn't always present, but where it was, it could comprise as much as a quarter of the mined ore material. In particular, iron was a key element in both bornite and chalcopyrite—two of the most abundant copper minerals found in Butte—in which it was chemically bound with the copper and sulfur in those minerals. While copper, iron, and sulfur combined to form two of the minerals, copper, arsenic, and sulfur combined to form enargite; and copper, antimony, and sulfur combined to form tennantite. Arsenic and antimony each comprised about 1 percent of the overall volume of the mined material. There were also occasional traces of manganese and zinc mixed in with the ores but never enough to generate a mineral formation with its own name.[3]

These mineral compounds represented a complicated problem for those who sought to exploit the material. Unlike gold and silver ore, which could be mechanically separated from quartz, the various elements were bound up with one another on an atomic level such that neither stamping alone nor roasting and stamping together could sever the chemical bonds. Excess sulfur in the sulfide ores could be coaxed out of the material through roasting, but it still left behind enough sulfur, as well as silicates and iron bound in the ore material, to continue to hold the copper in a chemically bound compound. The only method for releasing the copper completely was to smelt it, which meant replicating the extreme heat that had created the ore deposits in the first place but in an environment that allowed elemental iron, silica, and the remaining sulfur to oxidize and to either vaporize—as sulfur, arsenic, and antimony did—or coagulate in a floating slag, as iron and silica did, eventually leaving the copper behind in a nearly elemental form.[4]

Because of the nature of Butte's copper ores, the ore-processing plants built by the large copper companies in the 1880s and 1890s had included several generally similar stages of ore processing. Based on the quality and composition of the ore in a company's particular group of mines, its works would contain a combination of steps designed by smelting engineers to produce a smelting ore as quickly and cheaply as possible and a smelting procedure designed to purify copper according to the same ends. Throughout the 1880s and 1890s, the several large companies that had emerged in Butte had been fine-tuning these processes in their ore-processing plants based on the price of copper, the quality of material mined from their claims, and the ratio of investment to income from ore sales—which proved a tricky undertaking, given the changing circumstances of both the copper market and the ore lodes.[5]

Despite many significant differences in plant design across the smelting works and the use of various combinations of machinery and technologies, several broad strategies cut across the plants in the region and represented, in general terms, the necessary steps required to convert the various materials taken from the ore lodes into a refined and nearly pure copper product. The same processes were also ultimately integrated into the Washoe Reduction Works after consolidation.

The first step in the mass processing of ore was concentration, which was performed on all of the second-class ore, or about 90 percent of the ores taken from the Butte mines prior to 1907 and 100 percent after that. The concentrator was essentially a huge battery of stamps and an automated process of screens and gravity separation enclosed in a giant building, usually constructed on a hillside slope. The concentrating plants were designed to separate the valuable mineral sulfides from the granite and silicate and other non-metal-bearing materials that had been combined mechanically in the ores. Between 30 percent and 50 percent of the material dumped into the top of the concentrator left the plant as a pulverized waste known as *tailings*. Tailings dropped into a flume running underneath the concentrator and washed out of the plant in a steady flow of water to be deposited in a tailings pond. Because the goal of the concentrating process was an enriched roasting ore, more attention was given to the quality of the retained material than to the composition of the tailings—which resulted in a concentrating process that sent significant amounts of copper and other valuable and toxic metals, along with sulfur, into the tailings dump.[6]

The retained "concentrate" was shuttled from the concentrator to one of several kinds of roasting ovens installed in the plants across the district, where it was mixed with first-class ores and heated in a process called *calcining*. The goal of calcining, according to Edward Peters, was to heat the ores with "a comparatively moderate temperature with the purpose of effecting certain chemical, and rarely mechanical, changes required for subsequent treatment." The ores were heated in furnaces or steel cylinders with an indirect heat that would burn off a large share of the sulfur but hold on to as much iron and quartz as possible. By this method, the concentrate would be reduced from a material that contained about 30 percent sulfur to an oxidized "calcine" containing about 5 percent sulfur, with its iron and quartz material bound with oxygen atoms. The sulfur and trace amounts of the other metals were carried away as hot gases emitted out of the plant's smokestacks.[7]

The third step was the actual smelting, which involved using coke or charcoal mixed in with the calcine to initiate a heating process that would eventually melt the metals. This took place with the aid of forced air that used the oxidized iron and quartz as a fuel that superheated the material to higher and higher temperatures before floating to the top of the molten mass. The remaining sulfur and other metals (arsenic, antimony, manganese, and zinc) were expelled as gases as the material heated and melted. The molten iron and quartz were skimmed from the top of the molten copper and flumed to slag heaps. At the end of smelting, which took about eight hours per firing, a 95 percent to 98 percent "blister" copper remained. Until the electrical demand began requiring a purer copper, this copper product was loaded as it was in train cars and shipped east for sale. After the mid-1890s, the blister copper was treated one last time in a bath of electrified sulfuric acid in the West or in the East, where it was purified to better than 99 percent "electrolytic" copper before being rolled into wire.[8]

Trainload of ore for Washoe Smelter (undated). Massive amounts of rocky material were excavated from beneath Butte Hill and shuttled by rail to the mammoth new smelter in Anaconda, a facility that for years would represent the pinnacle in mass-production ore-processing techniques. Photograph by N. A. Forsyth; courtesy, Montana Historical Society Photography Archives, catalog #ST 001.175, Helena.

The measurement of copper output was impressive; millions of tons of material stretched across millions of miles of space, bringing electricity to all corners of urban domestic life. Equally striking were the unmeasured and unaccounted-for volumes of material that left the plant as waste. On average, across the district on any given day of production, for every five pounds of blister copper produced for sale, the plants produced ninety-five pounds of tailings, slag, and smoke—most of which was simply deposited in one form or another into the environment surrounding the plant. Almost half of this waste material, as seen earlier, was composed of pulverized granite and quartz tailings. Because of the crude concentrating technologies, tailings often contained high levels of copper and other heavy metals. Invariably, this powdery waste would not remain contained behind berms; it frequently washed downstream to be deposited elsewhere in the region. Another 20 percent of the material processed on any given day washed out of the plant as slag, an inert singed iron and silica oxide that had the appearance and texture of jet-black sand. Slag represented another disposal problem for plant owners, but because of its weight and texture, it rarely washed much farther than its disposal locations. The remaining 25 percent of the waste material left through the air in the form of smelter smoke composed of mostly (80%) sulfur now bound up with oxygen in a sulfur-dioxide gas and a small amount (20%) of arsenic, antimony, manganese, and zinc—now also bound up with oxygen as volatile oxides of heavy metal. Between the late 1880s and the early twentieth century, copper producers in and around Butte processed this ratio of material in their smelting plants.[9]

But the high percentage of waste product was only part of the story; during the same period, plant capacity increased exponentially such that in 1903 the single centralized Washoe Reduction Works processed eighty-one times as much ore as eight smelters together had produced in the early 1880s. In 1880 one ore-processing plant in Butte—William Clark and James Hill's Colorado and Montana—processed about 80 tons of ore a day, roughly 4 train cars of material. In 1885, four processing plants were running in Butte, among them processing as much as 450 tons of ore a day; at the southern end of the Deer Lodge Valley the Anaconda processing plant processed another 500 tons of ore a day, or just under 50 train cars of ore among them. In 1890, seven processing plants in Butte and two in Anaconda processed more than 3,900 tons of ore, or almost 200 rail cars of ore, a day—a threefold increase in just five years. By 1900, approximately 2,000 tons of ore a day were run through Butte concentrators, another 1,000 tons a day were processed in Great Falls, and the two Anaconda plants had been enlarged to a cumulative 4,000-tons-a-day capacity. Thus, in a mere twenty years, a small silver mining district had ballooned into the largest single copper-producing region in the world, processing the equivalent of 350 train cars of ore every day in a single processing location—as much ore every four days in the early twentieth century as had been processed in the entire year of 1880.[10]

It is impossible to know precisely how much of which kinds of waste were produced between 1880 and 1900 because of the copper producers' attention to marketable copper rather than to their waste products, but it is possible to generate estimates based on copper production levels and the average constituents of the Butte ores. According to such estimates, the Montana copper industry produced almost 2.5 billion pounds of copper in these years, which would have emitted almost 10 billion pounds of sulfur—mostly as a sulfur-dioxide gas—and just over a billion pounds each of arsenic and antimony, also in an oxidized and gaseous form. There would also have been about 25 billion pounds of pulverized tailings and roughly 10 billion pounds of slag material.

The patterns of exponential leaps in the scale of production made by the copper processors meant the distribution of these waste products was not spread evenly over time but came in exponentially larger batches, with most of the overall waste having been produced almost all at once after 1900. It took eleven years to generate one-quarter of the total volume of waste produced over the entire period but then only four years to double that amount and a mere six years to double it again. As much volume of waste was produced in the three-year period 1899–1901 as had been produced in the twelve years between 1880 and 1891, and as much waste was produced in the single year 1900 as had been produced during the first eight years of copper production in the Butte District. The expansive tendencies of the copper production system for Butte ores not only produced unbelievable volumes of material, but the numbers grew impressively larger each year.[11]

Most of the various waste materials found their way onto the ground within ten miles of their processing plant of origin, although a good proportion washed into creeks and rivers to be carried farther downstream. Through 1902, the area in which these growing volumes of waste material were disposed was spread among eight large plants concentrated in the Summit Valley, Great Falls, and Anaconda. After 1902, most of the processing of Butte ore became centralized at the mammoth Washoe Reduction Works in Anaconda as part of Amalgamated Trust's consolidation of Butte ore processing. The single large facility opened with a concentrating capacity of 6,000 tons of ore a day. By 1905, when Amalgamated finished closing its Butte smelters—the trust had purchased the Colorado and the Butte and Boston mines and processing plants in its bid to consolidate the Butte copper field a few years earlier—the Washoe Smelter had made plans to double its concentration capacity to 12,000 tons of ore a day, the equivalent of 600 train cars. Thus, by the middle of the first decade of the twentieth century, Amalgamated Trust was capable of producing 6,000 tons of tailings, 2,400 tons of slag, 1,200 tons of sulfur, and 600 tons of arsenic and antimony every day. In addition, this production of waste was no longer distributed across the mining district; it had become centralized into a single 250-acre facility at the southern end of the Deer Lodge Valley—a new $8 million plant investment that would have to work as efficiently as possible to derive a profit from the increasingly low-grade Butte ores now filling the company train cars. Mass production done on this scale had every chance of controlling the material flow process well enough to earn a profit, so long as nothing else stood in the way.[12]

"ACM Co." (smelter at Anaconda, aerial view, undated), Washoe Reduction Works. In 1903 this facility consolidated all of the ore processing from five copper smelters formerly located in Butte and bought out by Amalgamated Trust into a single mass-production copper plant. In 1906, following loud complaints from Deer Lodge Valley farmers, Amalgamated Trust constructed the enormous flume and 585-foot smokestack to capture particulate pollution before it left the factory. The arsenic and sulfuric acid poured from the plant day and night. Courtesy, Montana Historical Society Photography Archives, catalog #PAc 82–62.0070 AG, Helena.

FARMERS AND TAILINGS

On February 7, 1902, Butte attorney John Shelton filed papers on behalf of three Deer Lodge Valley farmers against each of four copper-processing plants, all Amalgamated Trust–owned companies with smelters in Butte. Each of the claims made the same charge: the waste products of the processing plants had destroyed the fertility of the plaintiffs' land and filled their watering ditches with harmful debris. The language of the complaints was almost identical, arguing that tailings from the smelter in question had washed into Silver Bow Creek and been carried into the Deer Lodge River, from which these farmers' irrigation ditches drew water. The debris had then settled into the bottom of the ditches, causing "the bed of the stream, as it is usually flowed across the land of said plaintiff, to be filled up and has diverted the course of the stream, and has caused the waters thereof to spread and overflow a large portion of the land." The complaint identified the offending material as "pulverized quartz and other debris in suspension, and copper, sulfur and other poisonous substances in solution," which these floods "deposited on the land." In effect, the farmers' ditches and agricultural fields had become large downstream settling ponds for the waste products of the ore-processing plants. Because the water carried and deposited materials that were "poisonous and injurious to all forms of vegetable and animal life," the upstream smelters had made it impossible for the farmers to irrigate their lands or successfully grow crops in the soils. Further, these impacts had not been limited to the plaintiffs' land, the complaint asserted; they had also ruined the waters of Silver Bow Creek and the Deer Lodge River. In fact, the complaint argued, "water for a distance of forty miles below said mill and smelter is thereby rendered almost useless for irrigation or any other useful purpose," by which impacts these plants constituted a "public nuisance." This included the entire stretch of river running through the Deer Lodge Valley. The complaints asked the court for an injunction against ore processing until the plants could devise a way to keep the tailings out of the waterway, and each also asked for damages for the negative impacts already caused.[13]

Cornelius Kelley, future president of the Anaconda Copper Company and at the time the local attorney for Amalgamated Trust, wrote the company's answers for each of the complaints. Like Shelton, he used nearly identical language in all of them. Kelley's argument was layered and complex, but it sought to convince the court that the charges had no merit. Kelley made no attempt to deny that the various copper-smelting facilities deteriorated the water quality. He admitted that each such facility produced tailings of the quality described in the complaints and that "the natural flow of the stream would carry such water and some portion of said tailings down the channel of the stream" but denied that adequate effluent had been emitted by the specific plant for which he was composing the answer to have filled the particular farmer's water ditch enough to cause it to overflow. Kelley's answers also admitted to contributing to the water pollution, but with the same caveat: "Defendant admits that its operations, as aforesaid, caused some pollution of the waters of the said stream; but it denies that the same became so polluted by any operation of this defendant as to become poisonous or injurious to all or any forms of vegetable or animal life." These statements on behalf of each of the four processing plants were a clever way to divide and dilute each individual plant's contribution to the overall water-quality issue and thus to complicate the legal responsibility for the cumulative impacts experienced downstream. But Kelley did not stop there.[14]

Each individual answer reasserted the right of each ore-processing company to use the waters for the purpose of ore processing. It then turned its sights on the actions of the plaintiffs. While the farmers had acquired their irrigation rights in the late 1870s and thus had made their argument based on prior appropriation rights to the use of clean water, Kelley answered that when the production facility in question had acquired its own right to use the water, the farmers had not objected. If the plaintiffs had objected before the mill was built, Kelley argued, the company would never have built its mill. Since it had done so without objections, the plant could not then be denied the water rights that came with the facility. Kelley argued similarly in each answer that the farmers were personally culpable in the damages to their lands by using water they knew was poisoned to irrigate their crops. Finally, each answer claimed that the economic hierarchies present in the Deer Lodge and Summit Valleys made the farmers' complaints oppose their own economic interests. The farmers' lands, Kelley argued, "owe their value for the most part to their proximity to the city of Butte which, during the period mentioned in the complaint, has steadily increased until there now exists in the city fifty-thousand people in the immediate vicinity of the smelting and reduction works." The only reason the farmers used the water and believed they had a cause of action in the first place was "because of the benefit derived by such owners of land on account of the direct and beneficial influence on the growth of said city arising from the operation of such reduction works." In this way, Butte "furnish[ed] the principal, and practically the only market for the product of such lands." For this reason, Kelley concluded, the farmers "ought not now be heard to complain on account of the alleged damage by reason of such use."[15]

The approach Shelton took in filing separate actions against each of the Butte ore production plants (and, curiously, not against the ore-processing plants in Anaconda, which were geographically closer to the farmers than the others and in the process of building a huge new facility) promised a series of duplicate cases over exactly the same matters of law, with a revolving set of plaintiffs and defendants. To streamline the cases, the judge asked the two sides to agree to a single hearing on the issue, pitting only one of the complainants against only one of the defendants. The parties agreed, and Shelton brought forward a single institutional client— Western Loan and Savings Company, which owned through default large acreage of farmland in the Deer Lodge Valley—to file the lawsuit against the Colusa-Parrot Mining and Smelting Company, the oldest copper processor in Butte. By that point, Colusa-Parrot was no longer processing ore, having been closed by Amalgamated Trust in 1900. The remaining cases were withdrawn pending the outcome of this lawsuit.[16]

The strategy did not pay off for the valley farmers. Both sides re-filed the same legal papers, and a short trial was held on evidence that appears to have amounted to photographs of the company's tailings at the mill and deposits of similar-looking tailings on the agricultural land. There was undoubtedly some testimony from both sides, although no transcript of the trial exists. After both sides presented their cases and rested, the judge gave the jury a set of instructions that effectively cut the central argument and the key damages entirely out of the case. Framing the case as requiring that the jury decide "between the alleged rights of mining and agricultural interests in the state," the judge asserted that "there is no way or method by which it can be ascertained how in what manner, or to what extent, the waters of said stream, or any thereof, have been affected or injured" by each of the five ore processors working along its banks or by the Butte sewage flow from so many residents. The only facts of law the jury could consider were the state of the Western Loan and Savings Company's land when the company purchased it in 1898 and whether that land had been damaged by the Colusa-Parrot Smelter after the smelter purchased the land. In case the implications of his constraints were not already clear, the judge also warned the jury against trying to assign damages to a single smelting company when several existed in the same watershed: "A farmer may be damaged, but unless he can prove satisfactorily that the person whom he sues has damaged him or contributed in a way to his damage, so that it can be intelligently ascertained in what proportion damage should be awarded, he could not recover [any compensation]." Unsurprisingly, the jury found the defendant, Colusa-Parrot, not guilty. None of the other cases ever went forward.[17]

In 1903, about a year after the massive Washoe Reduction Works began operations in Anaconda, Shelton filed a new water pollution lawsuit. Obviously having learned from his mistakes in the earlier cases, this time Shelton filed on behalf of a single claimant, farmer Hugh Magone, and named all of the companies running ore-processing plants in Butte and Anaconda as co-defendants. Magone had purchased farmland in the Deer Lodge Valley in the late 1880s and the 1890s, on which he had grown and bred a large herd of cattle as well as planted and harvested several market crops under irrigation from the Deer Lodge River. According to Magone's testimony, his output had grown fairly steadily throughout the 1890s, but around the turn of the century he noticed a decline in his crop yield. On closer inspection, he noticed that the farm's low-lying lands were covered with a layer of tailings, distinguished by a colorful shimmer in the light of day. By 1903, large areas of formerly productive farmland failed to yield any crops, and in other places the plants were sickly. As he had in the earlier lawsuit, Shelton identified the constituents of the tailings and pointed to tailings buildup in the irrigation canals and the deposition of poisons on irrigated lands as the primary damages inflicted on Magone's property by the ore production companies' water-use practices. The complaint asked for $20,000 in damages and an injunction against further operation by the ore processors until they could devise a method to prevent tailings from entering the water supply. (This case is discussed more fully later in the chapter.)[18]

In their answers, the smelters once again conceded that they had contributed pollution to the waters of the Deer Lodge Valley. Building from the instructions that had stymied the jury in the earlier water pollution case, however, the lawyers argued that it was impossible to establish precise ownership of the waste product emitted from separate plants. The Montana Ore Purchasing Company's answer put it succinctly. "It is not sufficient to show that it operated a concentrator from which copper might have escaped, and that copper is found upon the Magone ranch," Heinze's attorneys wrote, "but satisfactory proof must be found in the record that the Montana Ore Purchasing Company did contribute sufficient copper to in some way affect the productiveness of the Magone ranch. This nowhere appears." In each of the separate answers, the companies also claimed that they had initiated no significant change in their smelting activity since 1890. Kelley wrote again for all but Heinze's company, arguing that the smelters "have not at any time polluted the waters beyond the extent absolutely necessary in order to use the waters for the mining and milling purposes for which they have been used." If Magone's contention was upheld, Kelley argued, then the laws of Montana allowing water to be used in mining operations were meaningless: "an appropriation and use of water for the development and carrying on of Montana's greatest industry—mining—[could] not lawfully be made."[19]

Magone was not claiming that mining had no right to exist in the landscape but rather that sometime in the recent past it had begun to impact his prior right to use the same waters. While Shelton and his client, as agriculturalists, appeared rooted in an agrarian model of mixed husbandry agriculture, they did not in fact articulate common-law principles about the public good that inhered in flowing water. Instead, they claimed prior appropriation rights, the standard water law principle of the US West. In doing so, Magone was revealing, as many western farmers had done since the 1880s, how much agriculture had adopted mining institutions in constructing relationships with natural resources like water. Even for farmers in the early twentieth century, water was simply a commodity to be exploited. The mining ideal had become ubiquitous in natural resource use.

MINING AND WATER RIGHTS

PRIOR APPROPRIATION

Water makes ore processing possible. California gold rush miners had learned this fact quickly, and no credible hard-rock mining operation since the 1860s had neglected to consider the availability of running water when planning a mine and a mill. Water washed and sorted ores. It carried away tailings and slag. It is fair to say that the mountain west was built as a vast uncoordinated hydraulic system at the hands of mining *prior to* or concurrent with agriculture, but, as with agriculture, water was a requirement for its success. In the US West, water use and water policies grew up in the service of mining.[20]

Until the early rumblings of industrial-scale productivity in New England and the Northeast in the late 1830s, water use in the United States had been controlled by the riparian doctrine, a common-law tradition of water-use regulation in practice in England since at least the seventeenth century. The British courts, and the US courts following in their stead, had long ruled; and the riparian doctrine had long recognized that the use of water was a social good. The riparian doctrine treated water as a shared resource that existed for the benefit of all and thus could be owned by no one. The riparian doctrine had emerged alongside the institution of private property in land and as a strategy to regulate the social use of water under those conditions. In effect, the riparian right protected water, and thus water users, from the potential harm that could be done to it if it were treated as a private commodity in the service of a single interest. A person could own property *containing* a river or stream but not the water itself. Under the riparian doctrine, property owners were not prevented from using flowing water, but they had to use it in a reasonable manner and put it back where they found it as they found it. They were explicitly prevented from diminishing the quality of the moving water and were not allowed to remove the water permanently from its natural course. This legal principle accommodated multiple users upstream and downstream and reflected the natural resource ideals of a profit-oriented agricultural society.[21]

Following the British tradition, the United States had also treated rights to moving surface waters as separate from rights to property in land. One of the clearest and most frequently cited examples of the riparian doctrine was written into New York State water codes in the early nineteenth century. New York granted owners of private property full rights to all stationary water (lakes, ponds, and the like) and to any rainwater that fell on their land but not to creeks, rivers, or springs: "Water running in a definite stream formed by nature, over or under the surface, may be used by [the property owner] as long as it remains there." Further, property owners were not allowed to "prevent the natural flow of the stream, or of the natural spring from which it commences its definite course, nor pursue, nor pollute the same." In the first effort to clarify evolving water rights in the wake of the 1866 mining law, lawyer, judge, and legal scholar Gregory Yale wrote that the "riparian owners acquire no property in the water itself, but only the privilege of using it in its passage by reasonable interference." In most of the United States during the first half of the nineteenth century, no one could own the contents of a river; water was a public good that could be used but that had to be returned to its channel unblemished.[22]

As several scholars have shown, common law did not have an unproblematic existence in the ambitious and evolving United States. Throughout the nineteenth century, forces of commerce and industrialism gained social and economic traction through favorable legal changes, both in the laws themselves and in the general judicial style of ruling. Among the early victims of this shifting legal environment was the riparian doctrine, especially in cases of incongruous and mutually exclusive *types* of uses. In the Northeast, this situation pit textile manufacturers against upstream farmers and led courts to articulate new principles of water use and judicial activism. But beyond the centers of industry and their watersheds, riparian practices continued as a matter of course and were generally enlisted in the new western settlement. These practices began to end quickly everywhere after 1849, however.[23]

Gold-seeking pioneers had carried a great deal of cultural baggage with them when they made the long trek west across the continent in the 1850s and 1860s, but they did not bring the riparian doctrine. Instead, when organizing and codifying mining district rules, they sought expediency and efficiency, principles they had seen chipping away at the riparian doctrine in New England since the early 1840s. Gold miners applied the same principles to water that they had applied to placer claims: first in time meant first in right.

At the hands of placer miners in the context of placer gold mining, western water use evolved from the simple application of utility in a paying mining district to the acceptance of water as a commodity measured in mine feet. Placer miners needed to run water through their sluices and wash their pay dirt, which they then returned to the creek, usually temporarily laden with mud and debris. If the water was available for washing, the miners were allowed access; because they did not need clean water to wash pay dirt, there were no strong reactions to water quality. Sometimes, as happened to James Morley at Grasshopper Creek, upstream users would hold back so much water that there was less downstream than needed, but eventually the creek would flow again.

As time went on, mining districts, like those at Last Chance Gulch or in the upper reaches of Alder Gulch and Grasshopper Creek, were formed on creeks beds with insufficient water for washing. In these cases, water had to be flumed in from elsewhere. Building flumes from distant creeks or mountaintops and selling water to miners often represented one of the first large-scale industrial operations in placer gold regions. If one needed to wash pay dirt for gold, then water's natural place in the landscape was incidental; it could be treated as an accident (or, in some cases, as a cruel joke), but it need not remain in its original state in the channel if it was needed for work. Water's utilitarian role for placer gold miners generated unspoken and unwritten practices that were repeated throughout the mountain west during the gold rush era. Water was gradually treated as unfixed property. When it flowed untouched through a creek or a river or rested in a lake, it belonged to no one. But the moment a miner or ditch owner imagined a use for that water, he was welcome to use as much of it as he wanted, however he wished. At the hands of gold miners, water became privately accessible public property, valued first and foremost for its wealth-generating uses.[24]

Historian Donald Pisani has shown that these tacit principles first became codified in formal law when conflict arose between users. According to Pisani, the extension of property rights to water resources in the western placer mining districts was not new in itself but was simply an extension of pioneers' attitudes that had been developing since early in the nineteenth century. American pioneers had long believed that "property was created only when human beings produced something from the raw materials God had provided," Pisani wrote. Land that was not generating wealth was not considered property and, indeed, was often considered a waste of resources. As historian Elliott West pointed out, such an attitude about western lands made it easy for westering people to overlook the fact that they were appropriating otherwise occupied lands because of the general absence of wealth-generating activities among Native American groups.[25]

This utilitarian view of nature gave western water rights an entirely new flavor with the growth of fluming companies and the large investments that came with them. "During the 1850s, as corporations increasingly dominated mining," Pisani wrote, "the law changed from encouraging economic democracy to protecting capital." Conflict arose between corporate water companies and mining districts, both of which needed more and more of the scarce resource. "Only after mining passed from an activity engaged in by individual miners or miners organized in small groups to large hydraulic mining corporations—for which the miners worked as hired hands—did prior appropriation calcify into doctrine," Pisani concluded. Prior appropriation as the absolute right to property in water resources, to take and do what one wanted with the resource, only became necessary as interests larger than single claimants came to have an economic role in the placer districts. In most places, as in Montana, this occurred at about the time the gold rush peaked.[26]

While twentieth-century western historians revealed the layered power dynamics out of which both land as wealth and the doctrine of prior appropriation emerged, revealingly, nineteenth-century western judges and the legal scholars following their lead overlooked such nuances and tended to credit miners and mining district regulations with the animating role. "The proprietary right to the use of water depends upon the simple rule of prior appropriation like the right to a mining claim upon prior location," wrote Gregory Yale, the first legal scholar to compile mining jurisprudence in the late nineteenth century. "The extent of the [land] claim is limited by the district laws, while the quantity of water vesting in the owner is only limited by the purpose for which it was appropriated." Like the judges he studied, Yale ultimately viewed the mining district codes as having produced a new set of endemic legal principles that emerged almost organically out of the self-organized needs of the placer miners.[27]

Yale affirms Pisani's claim that prior appropriation became codified as companies fought with mining districts over water, but he suggests that the underlying reasoning of the courts derived from a view of district practices as a tacit common law among miners. The legal decision that set the precedent for prior appropriation in the US West emerged out of a lawsuit filed in 1855 by an organized mining district against a water fluming company that had appropriated all the water from the creek the district wanted to exploit. The placer miners had found potentially rich pay dirt and located a number of claims, but the creek bed no longer carried sufficient water to undertake placer mining operations. Well before the miners had located the mining district, another group of men had invested money in the construction of a dam farther upstream and a lengthy flume to convey the creek's water elsewhere for sale to other miners. The placer miners argued that the ditch company had violated their right to a free-flowing creek, their *riparian* right. They asked the court for an order to have the dam removed so they could initiate placer mining. The ditch owners argued, in turn, that they had established a legitimate business through legitimate means that would be damaged if the court granted the injunction. The court ruled in favor of the ditch company.[28]

In the written decision, it is possible to see how the new principle of water rights, while altogether new in American jurisprudence, followed the logic of mineral discovery: first in time, first in right. The case challenged the judges to construct a resolution to the existence of state laws that seemingly protected both parties. The miners "had a right to mine where they pleased," the judges wrote, but their location of choice occurred along "a stream from which the water had been already turned for the purpose of supplying the mines at another point." The court chose not to use the riparian doctrine to resolve the issue because doing so would have gone against state legislation. "However much the policy of the State, as indicated by her legislation, has conferred the privilege to work the mines," the judges continued, "it has equally conferred the right to divert the streams from their natural channels; and . . . these two rights stand upon equal footing." The conflicting rights of exploitation—the right to remove gold versus the right to remove the water—suggested to the judges that they evoked a different principle than reasonable temporary use. The court resolved the equal rights to conflicting use by applying the same principle to water as the mining district applied to gold: the principle of priority—"*que prior est in tempore potior est in jure* [who is first in time has the strongest legal claim]. The miner who selects a piece of ground must take it as he finds it." In this case, the right to appropriate the water was stronger than the right to mine because it was done first.[29]

While the *Irwin v. Philips* decision resolved a specific instance in which a group of miners had chosen to make claims along a dry creek bed, the court took it as an opportunity to extend the ruling to a set of general principles that would apply more broadly to water rights in California (and, ultimately, all of the mining west). Eleven years before the federal mining law would be passed in the US Congress, the California state court sought to affirm two legal principles on which the gold mining industry depended:

> If there are, as must be admitted, many things connected with this system which are crude and undigested, there are still some which a universal sense of necessity and propriety have [*sic*] so firmly fixed as that they have come to be looked upon as having the force and effect of *res adjudicata*. Among these, the most important are the rights of miners to be protected in their selected localities, and the rights of those who, by prior appropriation, have taken the waters from their natural beds, and by costly artificial works have conducted them for miles over mountains and ravines to supply the necessities of gold-diggers, and without which the most important interests of the mineral region would remain without development. So fully recognized have become these rights, that, without any specific legislation conferring them, they are alluded to and spoken of in various acts of the legislature in the same manner as if they were rights which had been vested by the most distinct expression of the will of the law-makers.[30]

In this way, the judges established the legal precedent that the mining district regulations had the standing of common law, the force of legislation. This did not preclude a different set of principles from the US Congress at a later date, but in the absence of federal legislation, the ruling secured some certainty. For Gregory Yale, *Irwin v. Philips* made clear "for the first time in any system of jurisprudence, that the right to the unlimited use of water in a running stream vested in the first appropriator, whether a riparian owner or not, with the correlative right to divert it to any extent, for sale or other use; and that subsequent locators, even for mining purposes, upon the banks of the same stream, as riparian owners, could only acquire an interest in the water for any purpose subordinate to the right of the first appropriator, provided any water was left."[31] For the first time in the history of Anglo-American jurisprudence and as a direct reaction to the new resource-use conflicts the practices of mining had created, water had become a commodity entirely separate from its natural channels and traditional flows, a material to be used by whomever acquired it first in whatever legal manner he pleased.

Within a few years, the California courts had forgotten the issues of law that had initially precipitated the *Irwin v. Philips* ruling and evoked the doctrine of prior appropriation as a natural product of the natural conditions of mining. In a famous case between miners, *Hill v. Smith*, in which the plaintiff evoked the riparian doctrine to seek damages caused by mining detritus in the river that interfered with his work, the judges ruled that the doctrine did not apply. First, the judges ruled that the principle was designed for private property; since the gold miners were working on public property, other standards should apply. The judge also claimed that the riparian doctrine was an agricultural principle; since mining and agriculture had fundamentally different needs from their natural landscape and as California was predominantly a mining region, more appropriate doctrines were necessary. "When the law declares that a riparian proprietor is entitled to have the water of a stream flow in its natural channel, *ubi currere solebat*, without diminution or alteration, it does so because its flow imparts fertility to the land, and because water in its pure state is indispensable for domestic uses," Justice Sanderson wrote in his decision. "But this rule is not applicable to miners and ditch-owners, simply because the conditions upon which it is founded do not exist in their case." Mining did not require that the water impart fertility, just solvency, so there were no longer economic or social reasons to require that it be returned in any quality or even at all. The decision rested squarely on the principle of prior appropriation rights. Mining as the foundational economic activity had priority to define natural resource–use policies, and the courts accepted this common practice.[32]

With the passage of the 1866 mining law, prior appropriation in mining regions became recognized by federal legislation. The 1872 mining law reiterated the same principle: prior appropriation would be the legal doctrine governing water rights in mineral districts and beyond in the US West. What had been an improvised practice designed to solve proximate difficulties in the absence of formal law thus became official policy without any formal legislative alteration.[33]

THE BALANCING DOCTRINE

For all its power to liberate water for economic purposes in California and the mining west, the principle of prior appropriation did not apply as easily in the East, where the tradition of riparian rights had long held sway in the adjudication of water rights. Growing conflicts, however, seem to have encouraged eastern judges to construct new principles by which to adapt legal decision-making between competitive users under the riparian doctrine. Judges turned their attention to the "reasonable use" clause and began to re-conceive the meaning of the term in the context of the expanding and changing economic practices in the landscape during industrialization. In an industrializing nation, they reasoned, there were many more "reasonable uses" than those imagined by British and American farmers when these water rights were framed. Beginning in 1870 in Pennsylvania, the preeminent East Coast

mining state, new patterns of legal reasoning began to consider mining activities—
at least initially thought of as the industrial production of human necessities—as
uses of water included in the category *reasonable*. As a "reasonable use" of water,
these industrial activities were thus legally identified as public goods, giving them
legal rights superior to individual rights. During a period when lode mining began
its iterative expansion in the West and coal mining increased everywhere coal could
be found, these cases began to chip away at the already weakened principle of riparian rights.[34]

The changes were incremental and occurred as a gradual transformation of underlying legal principles, but their overall impact and the ultimate new principle of legal
reasoning they generated can be seen in the sequential rulings of a water rights case
in eastern Pennsylvania. The case, *Sanderson v. Pennsylvania Coal Company* (1872–
86), pit an individual property owner against a large coal mining company. In the
early 1870s, Sanderson had purchased land along Meadow Brook outside Scranton,
Pennsylvania, to build a house. To supply the property with adequate water, he channeled water from the creek—which ran through his property—into a holding pond.
Sanderson pumped water from the pond through an elaborate system of piping and
pumping engines into a large cistern atop his house for general household use, and
in the winter he cut ice from the pond. Sanderson also stocked the pond with fish
for his consumption. Shortly after Sanderson completed his suburban paradise, the
Pennsylvania Coal Company opened a coal mine near the creek, about two miles
upstream. Within months of the initiation of mining, Sanderson's fish began to die,
as did the willows that surrounded his pond. Sanderson also discovered that the pipes
that drew the water from the stream were severely corroded. Soon, the water was vis-
ibly polluted and too diminished in quality to be of any worth for household or other
purposes. The polluted and corrosive water in his pond forced Sanderson to abandon
his country retreat. He sued the mining company for damaging his property.[35]

The impacts on the biota and on Sanderson's iron pipes were the result of the
highly acidified water pumped out of the coal seam by the coal mining company
and discharged into the river. The acidity came from the interaction of water and
the organic compounds in coal, precipitating sulfuric acid in the breakdown of the
organic chemicals. To gain access to saturated coal seams, coal companies, like metal
ore mining companies, had to extract the water from the underground workspace.
The common practice in both eastern and western mining regions was to drain the
mines directly into nearby rivers, where it was believed the acidic mine waters would
be diluted and eventually dispersed into the ocean. In this instance, the volume of
mine waters had clearly overwhelmed the volume of the creek as well as Sanderson's
pond, leading to the various impacts he experienced. By the standards of the riparian doctrine, Sanderson's water rights had been violated by the actions; it was on this

basis that he sued the Pennsylvania Coal Company, seeking nuisance damages and an injunction against disposing mine water directly into the creek.

In its defense, Pennsylvania Coal claimed it had operated entirely within its rights and had thus committed no violation. It had undertaken a lawful mining operation in which it did nothing illegal or out of the ordinary. Part of this lawful operation, the company argued, necessarily included removing certain amounts of water from its mines; if it could not remove that water, mining would cease. Mine water, Pennsylvania Coal argued, was a natural by-product of coal mining, and it was merely a natural accident that the coal happened to be near the creek and an accident of location that extracting the coal led to the pollution of the water in the creek. The company asked the courts to recognize the reasonable needs of important industries such as coal mining and to consider their importance to the health of the nation's economy when they ruled on the use of common resources like water.[36] The company went so far as to suggest that the work of coal mining and its unavoidable effluents were perhaps even more natural than Sanderson's fish pond, which Pennsylvania Coal was charged with illegally damaging: "If, in our case, we cannot use our land for the natural purpose of mining coal because our neighbor cannot keep his tame fish in his pond, the same rule should apply to him. He should not be allowed to maintain a fish-pond so near our mine that we cannot use it."[37]

Sanderson v. Pennsylvania Coal returned to the Pennsylvania courts four times between 1873 and 1886, tracking the unsettled and changing jurisprudence around water rights. Initially, the Pennsylvania state court dismissed the case as a "non-suit," ruling that Sanderson had not met the evidentiary standard for his claim against the company. That is, he had not proven that the coal company's actions constituted a violation of his water rights. Sanderson then took the ruling to the appellate court, which reversed the lower court's decision. There was no doubt, the appellate court reasoned, that Pennsylvania Coal had undertaken a lawful activity, "but however laudable an industry may be, its managers are still subject to the rule that their property cannot be so used as to inflict injury on the property of their neighbors." When Pennsylvania Coal appealed again in 1880, the courts reached a similar conclusion. Thus, despite the state court's preliminary refusal to hear Sanderson's case, throughout the 1870s the underlying social considerations embedded in the riparian doctrine continued to inform the courts' decisions.[38]

In 1883, Sanderson finally received a favorable judgment and was granted a large cash award by a third court, but the Pennsylvania Coal Company would not let the judgment rest and appealed the decision. A new group of appellate judges reversed the ruling against the company and in the process articulated a new basis for adjudicating water rights. In its appeal, Pennsylvania Coal had moved to reopen the question of liability, in part to attempt to lower the damages levied against it. But the company also raised the question of its water rights once again. The strategy was more successful than the company could have hoped. In 1886 the Pennsylvania appellate court articulated a fundamentally new line of reasoning to adjudicate nuisance lawsuits:

> The plaintiff's [Sanderson's] grievance is for a mere personal inconvenience, and we are of opinion that mere private personal inconvenience, arising in this way and under such circumstances, must yield to the necessities of a great public industry, which, although in the hands of a private corporation, subserves a great public interest.
> To encourage the development of the great natural resources of a country, trifling inconveniences to particular persons must sometimes give way to the necessities of a great community.[39]

The justices attempted to temper this view by clarifying that they did not mean to say that an industry could enter private property and inflict harm on the owner or the owner's land, but the logic was established and what would be called the *balancing doctrine* was born. In effect, in cases where two parties claimed conflicting rights on a shared resource, what had once been a question about the quality in which the resource was left after being used now became a question of the social value of the user, a value the courts who used this doctrine measured in purely economic terms.[40]

While the balancing doctrine created an altogether new means for courts to evaluate competing rights, one that implicitly valued the protection of capital over other traditional social values and elevated the economic value represented by different sides of a lawsuit as a relevant issue in the determination of mining rights, legal scholars at the time believed it was merely the logical transformation of common law to a new set of geographic circumstances. According to Curtis Lindley, the balancing doctrine made as much sense in the US mining landscape as the riparian doctrine had made in Great Britain: "The mines in England are generally located in highly improved sections, where the land possesses great intrinsic value and the streams are filled with choice fish, the sole right to which is in the nobility and landed gentry. Under such circumstances, we could hardly expect the English judges to lay down a rule suited to the rough mountain lands which, in the main, constitute the mining regions of Pennsylvania."[41]

While the balancing doctrine became a useful legal tool as coal mining expanded across Pennsylvania, it did not find immediate traction in the western courts. In certain respects, the slower acceptance of the eastern principle can be attributed to the existence of prior appropriation as a principle of western water law. Its existence, in effect, would have forestalled the need for an alternative reasoning for the "reasonable use" principle of riparian rights. However, it was evoked now and again in conflicts over shared resources. When it was, the judicial push-back was swift and concise, suggesting that its implied economic principles were antithetical to national economic development in a democracy.

The most succinct statement of this form of opposition was articulated by Ninth Circuit federal judge Lorenzo Sawyer in 1883. In his ruling reversing a California decision that had used the doctrine of "superior interest" to rule in favor of a large plaintiff on the basis of its economic size, Justice Sawyer argued that such reasoning not only abused the purposes of law, but it would create circumstances undesired in a democracy. "It is by protecting the most humble in his small estate against the encroachments of large capital and large interests that the poor man is ultimately enabled to become a capitalist himself," Sawyer wrote. "If the smaller interest must yield to the larger, all small property rights, and all smaller and less important enterprises, industries, and pursuits would sooner or later be absorbed by the large, more powerful few; and their development to a condition of great value and importance, both to the individual and the public, would be arrested in its incipiency."[42]

Sawyer's reasoning continued to hold sway two decades later when farmers in Utah sued a nearby ore processor for the impacts of its effluents on farmland. In the 1904 case *McCleery* v. *Highland Boy Gold Mining Co.*, Ninth Circuit judge John Marshall ruled on an appeal by the mining company. The Utah district courts had issued an injunction against the company and levied damages for its impacts on the plaintiff's land. The Highland Company appealed the ruling, claiming that the injunction had a significant effect on the region's economic health. The company argued that it should be allowed to continue ore processing as long as it paid the injured parties for their losses. "If correct," Justice Marshall stated in response, "the property of the poor is held by uncertain tenure, and the constitutional provisions forbidding the taking of property for private use would be of no avail."[43] He continued:

> As a substitute it would be declared that private property is held on the condition that it may be taken by any person who can make a more profitable use of it, provided that such person shall be answerable in damages to the former owner for his injury. In a state of society the rights of the individual must to some extent be sacrificed to the rights of the social body; but this does not warrant the forcible taking of property from a man of small means to give it to the wealthy man, on the ground that the public will be indirectly advantaged by the greater activity of the capitalist. Public policy, I think, is more concerned in the protection of individual rights than in the profits to inure to individuals by the invasion of those rights.[44]

Courts continued to uphold injunctions in both Utah and Arizona during this period, with both state and federal courts agreeing that the balancing doctrine was a dangerous principle and a precedent unworthy of the American legal system. Despite the western courts' resistance in these cases, however, large interests in the

West—especially those financed by men in the East, as was Amalgamated Trust—continued to evoke the principle with the ambition that at some point it would eventually stick.

MAGONE VERSUS THE COPPER INTERESTS

Thus, in the wake of two countervailing legal trends connected to the use of public natural resources in the West, John Shelton brought a lawsuit (discussed earlier) on behalf of farmer Hugh Magone against the nation's largest collection of copper companies. On the one hand, the principle of necessary use embedded in the doctrine of prior appropriation seemed to give the ore processors a strong legal argument that they had a right to use the waters of the Summit and Deer Lodge Valleys in whatever way they needed to for their legitimate business. On the other hand, Magone had purchased land with associated agricultural water rights and had practiced agriculture successfully for more than a decade when the tailings had begun to have obvious impacts on his property. The general disdain for the balancing doctrine—invoked by the companies in the earlier lawsuit brought by Western Loan and Savings but not included in the instructions to the jury—seemed to favor Magone's economic rights and to point toward a potentially successful financial judgment, long-term relief from water pollution that would follow an injunction, and more effective tailings containment by the processors.

After the initial complaint and answers, which raised most of the same issues the Western Loan and Savings case had raised, three years of discovery and court filings followed as each side argued over principles of law and slowly assembled the evidence and arguments for its case. By 1906, the court docket was huge. Not only had Shelton gathered production statistics from the ore processors, but he had also had scientists complete studies on the effects of waste products, subjecting each of a number of important Deer Lodge Valley agricultural plants to the various materials emitted by the processors. He had also had photographs taken. Because of the mass of evidence and expert reports, district judge William Henry Hunt referred the case to a special judicial magistrate, Oliver T. Crane, who he suggested would have both the time and the undivided attention to weigh all of the evidence and arrive at a decision.[45]

The trial itself began in late 1905. Shelton's witnesses ranged from farmers in the Deer Lodge Valley—who testified to the diminished yields they had experienced in the wake of mining expansion—to plant, water, and soil experts from Montana Agricultural College in Bozeman, who identified specific ore reduction by-products, their presence in the water, and their known effects on farmland and crops. To establish that the cause of the damage could be found in the smelter residue, Shelton presented a series of plant studies done by water and plant specialists at Montana

Agricultural College that showed the effects of each of the various chemicals emitted from the smelters in Butte on the plant life in the Deer Lodge Valley.

The farmers also provided compelling testimony during the trial. One farmer testified: "In some places on our premises the ground is bare entirely. The sediment has accumulated to such an extent that nothing grows. There are other places where there is a fairly good growth of grain or hay, but the fields become very much spotted. Where the water runs for some length of time the sediment becomes so thick that nothing will grow, and in our mode of irrigation where it has only received slight application, the crops are growing to some extent, but not so good as in former years." Another farmer testified that in the late 1890s, he and his partners "began to discover that there were places in our fields where the crops were lighter, and the vegetation did not have a healthy appearance. The plants began to assume a sickly pale color and as the season progresses some of them die, and some fail to mature." While defense attorneys objected every step of the way and greatly extenuated the farmers' testimony, Shelton's strategy was to show that Magone was but one of many farmers in the Deer Lodge Valley who had watched the irrigation waters decline in quality during the 1890s and eventually become the source of damage to their crops and farmland. Through visual, experiential, and scientific evidence and testimony, he sought to make a clear connection between these impacts and the effluents from the defendants' processing plants. Magone could not afford to pay the expert witnesses to show up for the trial, but their reports were submitted to the court as evidence.[46]

The defense, which sought to undermine Shelton's seemingly clear-cut case of negligence and damages through several strategies, offered testimony from its own farmers, all of whom declared that they had experienced no such decrease in fertility on their lands. The defense attorneys developed the theory that the real impact on the soil fertility had resulted from improper irrigation techniques, causing the precipitation of alkalai to the surface, a theory supported by the testimony of Deer Lodge ranchers. The defense also brought processing-plant engineers who described concentration practices at several of the mills, assuring the court that the best and most careful practices were followed. The company even rounded up a soil expert who suggested that even if some acidic tailing material had made its way to Magone's ranch, the base alkalai would have reacted with it to render it harmless.[47]

The farmers and expert witnesses presented by the defendants were enlisted not only to cast doubt on the certainty that the copper processors' effluents had impacted Magone's ranch but also to support the copper companies' broader contention: they had a right to pollute these waters, and nothing they had done as operators had exceeded that right. According to the legal documents filed by company attorneys and the closing statements of the defense, they had acquired the right to use the waters of Silver Bow and Warm Springs Creeks through prior appropriation.

Bundled into such a right was the ability to use these waters as they needed to perform the activities for which they had appropriated the water. Such legitimate uses included the washing and sluicing of ores, tailings, and slag. The incidental pollution of water through this legitimate use by the mining industry did not constitute negligence.[48]

The companies bolstered the claim to responsible work by arguing that they had engaged in essentially the same practices in their plants since ore processing began in the 1880s. While accurate on its face, such a claim glossed over the substantial changes in the scale of output during the same period, which would have been experienced as the qualitative change testified to by Magone. The companies suggested that rather than their malfeasance, Magone was an opportunist who had frivolously brought this lawsuit—in the words of one Montana Ore Purchasing Company attorney—to "raise a crop of cash from the various corporations, rather than a crop of cereal from the land." In an exasperated tone, the attorney asked rhetorically, "Can it be that these concentrators have contributed for fifteen years toward poisoning this stream, and the poisonous substance cannot be detected until within the last few years?" While a close look at the changes in scale would suggest that the answer was a certain "yes," the finding of facts by the special magistrate led Judge Hunt to take his ruling in the lawsuit in the opposite direction, even beyond the arguments made by the defendants.[49]

One year before the final ruling in the case, the Ninth Circuit Court rendered a decision in a lawsuit brought by farmers against a mining company in nearby Kellogg, Idaho—*McCarthy v. Bunker Hill & Sullivan Mining & Smelting Company*—in which, for the first time in the West, the federal courts evoked the balancing doctrine in their decision. As in the Magone case in Montana, Idaho farmers were complaining about an ore-processing company dumping tailings into the river and sued for an injunction against the metal producer. Ninth Circuit Court judge James Beatty, writing for the majority in the decision, acknowledged that many farmers were impacted by the smelter waste and that the Bunker Hill Smelter was causing the damage, which was extensive. An injunction on the smelter, however, would require closing not only the mill but the mines as well because the district's low-grade ores could not be profitably shipped elsewhere for treatment. In weighing the prudence of issuing an injunction against the Bunker Hill company on behalf of the farmers, the judge had to weigh the economic impacts:

> The court must consider the consequences of closing the mills and mines. It must
> bear in mind the great hardship and loss to the defendant. They have millions of
> dollars invested in their properties and are now conducting an immense business,
> which is not only of much profit to them, but also of great business interest to others.

But of equal consideration is the fact that it would deprive thousands of laborers of employment who are now earning good wages; also there are many others engaged in various avocations who would be seriously affected. I presume it is safe to say that there are ten thousand to twelve thousand people who are now earning a livelihood through the operation of these mines and mills, all of whom would be seriously injured by an injunction. The court will long hesitate before taking such a drastic mode of guarding complainants' interest, as would result in incalculable injury, not only to defendants, but also to large communities.[50]

Despite Sawyer's and Marshall's strong admonitions against the use of the balancing doctrine in settling such cases, Beatty determined that the rights of smaller, lesser economic interests—such as farmers—could effectively be trampled as long as some form of compensation was forthcoming.

When the Magone case was finally decided, Judge Hunt made direct reference to the principles outlined in Beatty's 1906 ruling, denying the injunction on the basis of the balancing doctrine. When the copper-processing companies had appropriated waters from Silver Bow Creek during the 1880s (several years prior to Magone's purchase of his Deer Lodge ranch), they had acquired the prior appropriation right to ongoing use for their business purposes. "Included in the enjoyment of the right," the justice ruled, "there is necessarily included the right of pollution of the water to some extent, provided the use makes pollution necessary—very slight perhaps if the use is for irrigation, oftentimes comparatively serious if the use be for mining or smelting." For this reason, when Magone acquired his rights to water use in the early 1890s, he became a secondary appropriator who could not expect the companies to pollute any less than they had at the time he acquired the right. It followed, then, that Magone could not "demand that the water in the stream be as pure as he would have it for irrigation." Judge Hunt ruled that, given the existing water pollution in the rivers when Magone acquired the secondary right, he could not recover any damages for impacts to his crops or fields that had resulted from his own irrigation practices, which represented the majority of his original estimate of $20,000 in damages. Instead, the only claim Magone had against the copper processors was for the deposition of tailings onto his property that had resulted from the flooding of the ditches, amounting to just under $2,000 in damages. As historian Fred Quivik characterized it, "The companies received little more than a slap on their wrists."[51]

BLISS VERSUS AMALGAMATED TRUST

The balancing doctrine would be invoked one last time to protect Amalgamated Trust's largest holding in Montana—the Anaconda Copper Company—from lawsuits filed against it by the Deer Lodge Farmers' Association, an organization of valley farmers. In 1902 the company had built and then moved all of its copper processing into the Washoe Reduction Works at the southern edge of Anaconda—across the Warm Springs Valley from, and a few miles closer to Butte than, the Upper and Lower Works. Almost immediately, farmers in the Deer Lodge Valley had begun to notice negative impacts on their farm animals. A group of them brought this to the attention of the company, which offered to pay damages ($330,000 in all) and began constructing a huge flue and smokestack connecting all of the roasters and smelters in the enormous facility to a single point of emissions out of a 500-foot-tall smokestack atop the hill above the processing plant. The company claimed that emissions at this height, 1,100 feet above the valley floor, would dilute any particulate matter that made it up that high and reduce the impacts on the valley ranches. While the processing-plant managers claimed publicly that these additions were constructed to protect valley farmers from any smoke emissions, the fact that the company added a profitable arsenic production facility in 1903 and quickly began selling this toxic by-product to pesticide manufacturers suggests that it had more than one goal for the enormous flue. The way the company responded to subsequent complaints by valley farmers further supports skepticism that the flue and smokestack reflected a genuine interest by the company to be a good neighbor and adds weight to the farmers' suspicion that the flue and stack had always been planned by the company and had merely been turned into a smoke abatement technology in the midst of the original complaints.[52]

In 1904, a year after the flue and smokestack had been added to the Washoe Reduction Works, a group of about 100 farmers met in the Willow Glen schoolhouse outside Anaconda to compare notes and determine how to contend with what they agreed was the continued negative impact of the smoke on their crops and livestock. Tallying up their individual cases, they determined that with the ill health and deaths of some of their animals, the impacts on yields from their fields, and the value of their collective property affected by the toxic smoke, they had incurred $2 million in damages. As farmers had done in 1902, they brought their assessment and complaint to the manager of the Washoe plant, hoping to be compensated for their damages and to sell their lands to the company so they could take their farming operations somewhere else in Montana, where the risk of pollution did not exist. They notified the company that they needed an answer before the next growing season, setting a May 1, 1905, deadline for an answer. By late winter 1905, they still had received no response. When they inquired about the company's reaction, they were told that any attempt at a lawsuit would be their financial ruin.[53]

The farmers decided that their only choice was to develop a case against the company and take it to court. They raised $40,000 and convinced former Butte and then

Panorama view of Butte from the southwest, circa 1895. With the consolidation of ore-processing operations into the Washoe plant in Anaconda, the total load of smoke that riddled Butte's air in the mid-1890s began to flow through a single masonry chimney in the southern Deer Lodge Valley. Unknown photographer. Courtesy, Montana Historical Society Photography Archives, Tilton Collection, catalog #946–031, Helena.

Idaho resident Fred Bliss to act as their plaintiff. On May 5, 1905, the Deer Lodge Farmers' Association filed a complaint in federal district court. They claimed that since 1902, when the company had initiated operations at the Washoe Reduction Works, 60,000 acres in what the complaint called the "smoke zone" surrounding the plant had been impacted and in many cases utterly destroyed by the sulfur and arsenic fumes spewing out of the company's smokestack. The association did not believe it could get a fair trial in the local Butte courts, so it recruited Bliss, who as a resident of Idaho and the owner of 320 acres within the smoke zone created the cross-jurisdictional claim that made this a federal cause of action against the local smelter. The lawsuit declared the copper-processing facility a public nuisance and requested that the courts issue an injunction against further copper ore processing until methods could be found to prevent emissions into the predominantly agricultural valley.[54]

Cornelius Kelley, who was simultaneously acting as lead attorney on the Magone case before the same federal judge, wrote the answer for the defense, using the company's recent actions with farmers and the flue and smokestack construction as a launching point for a wide-ranging counterargument. The company never believed the Washoe plant had done any substantial damage, Kelley wrote, but in good faith to its neighboring farmers, it had not only paid significant damages in cash compensation but had also done everything reasonable to abate even the perception of smoke impacts by constructing a $750,000 flue and smokestack through which significant proportions of gaseous solids were settled out of the smoke and out of which any small amounts of remaining solids that might be emitted were sent so far aloft in the atmosphere as to be completely diluted before they could reach the ground. The answer challenged the farmers' depiction of the Deer Lodge Valley as valuable for agriculture and repeated the claim in the Magone case that the soils were overly alkaline, thin, and not very fertile. Based on all this, Kelley argued that the farmers' estimate of the land value in the region affected by smelter smoke was four times too high.

The answer also challenged the claim that the company had somehow begun something new in 1902, arguing that it had been almost continuously processing Butte's low-grade copper ore in Anaconda since 1884, and never before had anyone complained. Kelley admitted that these low-grade ores did contain a certain amount of sulfur and arsenic, but they were a natural by-product of the legitimate business in which the company was engaged. Finally, Kelley argued that an injunction against the company's operation would have broad and deep economic impacts on two of the largest cities in Montana, Butte and Anaconda, whose combined population of 70,000 people depended on the operation of the ore-processing plant. More than $11 million in wages and other company spending in the state would evaporate. An injunction would also impact the nation's and the world's copper supplies, as the Washoe plant supplied 20 percent of the former and 10 percent of the latter. In effect, Kelley asserted, this copper-processing plant rested in the midst of a web of economic activities too big to enjoin.[55]

Over the next three years, both sides hired experts to perform investigations and studies of the valley, undertook full-scale and costly trials before a dedicated special magistrate (Judge Hunt again felt his time was better spent doing other things), and then—after producing thousands of evidentiary exhibits and more than 25,000 pages of trial material—waited anxiously for the judge's decision. The plaintiffs generated studies establishing that the new stack emitted at least 44,000 pounds of arsenic trioxide (the gaseous and highly volatile form of the heavy metal) a day, but they ran out of money for a study of sulfur emissions and thus generated no estimate. Their experts did perform biological studies and determined that the lesions, internal organ failure, and developmental anomalies experienced by livestock in the valley were caused by arsenic. They also performed studies that detected arsenic in the soils and the tissues of plants as much as thirty miles from the smokestack. The defense, which had much deeper pockets and brought in chemists and veterinarian experts with national reputations, generated studies of its own—mostly a series of mass dissections in which the experts claimed that any lesions or organ deterioration found in livestock had been caused by parasites infecting the valley. The experts also took pains to question the abilities and research skills of the plaintiff's experts.[56]

The Deer Lodge Farmers' Association had no idea what it was getting into and quickly spent all of its $40,000 in trial funds. The association scrambled, turning to the findings of state veterinarian investigations, studies done by US Forest Service scientists (the Washoe plant was nestled against a national forest to its south), and experts from the US Department of Agriculture (USDA)—all of whom, incidentally, provided additional strong support for the contention that the smoke was having damaging impacts on livestock and crops. The copper company, in contrast, boarded its experts—several of whom were former McGill University classmates of Washoe plant superintendent E. D. Mathewson—at the luxurious Montana Hotel in downtown Anaconda and was quick to pay any expenses incurred in its experts' studies, such as the cost of 100 valley cattle slaughtered in the autopsies that claimed to have uncovered the mysterious parasite.[57]

In late 1908 the special magistrate in the case submitted his findings of facts to Judge Hunt, who rendered his decision the following May. The findings and logic of the decision were extensive and detailed but can be reduced to several key conclusions. The court determined that the company had indeed operated its business in Anaconda for twenty continuous years, treating the same generally low-grade ore during the entire period, which necessarily included the production of sulfur and arsenic waste. It found that the renovations and reconstructions that went into building the Washoe Reduction Works, especially the geographic shift down Warm Spring Valley, had created temporary increases in the impacts of both sulfur and arsenic compounds on livestock and plants but that the addition of the flue and smokestack in 1903 had ended the damages and injury caused by sulfur compounds. Arsenic, the court determined, was a different story. The court found that arsenic had indeed continued to be emitted into the valley from the company smokestack and to be carried into the valley through the air, "depositing at times sufficient quantities of said arsenic on the hay, grasses, and fodders thereon to injuriously affect and poison many of the livestock which eat said hay, grass, and fodder." Further, the court found that such depositions rendered these farmlands "less profitable and less valuable." Based on the court's estimation, it believed the plaintiff, Fred Bliss, who had tried several times to lease his 320 acres, had as a result of the arsenic suffered $350 in damages. The court noted that the dispersal of arsenic throughout the southern Deer Lodge Valley was impacting the large number of ranches there, but none of them had filed suit in this case, and none of them would have had standing in the federal case anyway.[58]

Yet while the evidence mustered by the underfunded Deer Lodge Farmers' Association had clearly held sway with the court (there was nary a mention of parasites anywhere in the finding of facts), Justice Hunt refused to grant an injunction against the processing plant. Arsenic or no arsenic, Hunt reasoned, the handful of farmers whose operations experienced some negative effects from the smoke paled in economic comparison to the large interest the company—which depended on its copper-processing plant—had created in the state. He mentioned the city of Anaconda, which the company had built on "vacant and unoccupied" lands and elevated to an assessed value of more than $3 million—all of which depended on the continued operation of ore processing—and whose occupants had offered a "ready market and good prices" to the Deer Lodge farmers. Hunt pointed out that the company had contributed almost half of Deer Lodge County's property taxes. He pointed to Butte, where company operations contributed almost 30 percent of the taxes in Silver Bow County and a population of 70,000 "depends upon the continued operation of the copper mines, and about two-thirds of all ores mined at Butte are treated and reduced" in the Washoe plant. He pointed to the "vast quantities of coal, coke, and lumber" the company purchased in Montana, amounting to the expenditure of more than $17 million in the state since 1902. Hunt mentioned the more than 500 million pounds of copper produced from the new plant between 1902 and 1906, representing 20 percent of all copper produced in the United States in those years. A large private interest with such extensive social dependencies could not be asked to cease its operations for anyone.[59] The balancing doctrine gave Hunt all the reasoning he needed.

The Deer Lodge Farmers' Association appealed the decision up through the federal courts, losing at each step and then failing to get an audience for the case before the US Supreme Court because it no longer had the money even to cover the requisite filing fees. The studies of arsenic poisoning done by the USDA veterinarian were published in the years ahead and became model research in epidemiology. The national experts who developed the parasite theory, by contrast, never published a sentence of their work, despite being academic scientists whose reputations were presumably made through published research. But evidence of poisoning, the impacts on farmers, and the potential long-term damage to the Deer Lodge Valley and beyond were simply no match for a legal doctrine that acknowledged a right to pollute shared resources and weighed conflicting private property rights on the basis of the litigants' economic power—a legal doctrine forged wholly by the needs of metal mining.[60]

PREVENTING POLLUTION IN A MINING SOCIETY

The nature of Butte's copper ores had created three decades of expansive turmoil by 1910. Struggles by the early capitalists to keep up with their investments led to rapid expansion of copper operations and quickly squeezed the small operators and those without access to a steady flow of capital out of the field. In the process, the overproduction created its own market justification, with a growing spread of electrification and market demand that nearly kept pace with production. As deeper mines and larger interests generated new conflicts around the uncertainty of owning a buried resource that tended to wander under other people's property, the move to consolidate the Butte mines under a single management became an increasingly appealing and apparently rational strategy for the district. The plan was nearly completed by the dawn of the twentieth century.

The nature of the copper production system, shaped by the nature of Butte ores, continued to complicate the original goal of investors, which was to build a system whose investments stayed ahead of its costs. Uncertainty had troubled the system from the start, and just as consolidation got under way, new uncertainties arose. The exponential growth in copper production and its concurrent exponential growth in waste production, coupled with Amalgamated Trust's decision to centralize all of its ore processing into a single new facility in Anaconda, proved to be one step too far for the farmers of the Deer Lodge Valley and the health of their assets in livestock and land. They suddenly rose up in opposition to the new, polluted condition of their air and water. Fortunately for the copper trust, however, mining practices had been generating incremental changes in the adjudication of water rights and the legal standards of reasoning in public nuisance lawsuits. In the same way Amalgamated Trust had gathered the growing resources and production of the various copper interests in Butte to create a sudden transformation of the Deer Lodge Valley environment, federal courts in the West had been increasingly shifting in the way judges valued economic growth more than other, more traditional rights like clean water and individual private property. This shift transformed traditional rights into the sudden legal sanction of a new scale and scope of natural resource devastation. In an unintended confluence of new environmental impacts and legal standards, mining succeeded in creating its own legal justification at the very historical moment its social relationship with nature became dangerously unsustainable in the United States.

Both environmental scholarship and modern environmentalists have long claimed that activities in the twentieth century impacted human health and biological systems because scientists and engineers, never mind farmers and other, more ordinary citizens, did not understand the dangers posed by the new technologies and materials. The United States had to learn from its mistakes, this claim argues, and those mistakes only became apparent in the late twentieth century when they had become broad enough and deep enough that they could no longer be ignored.

But the dangers faced by the known and unknown hazards and toxins pouring out of the Washoe Reduction Works into the Deer Lodge Valley were as clear as day—not only to valley farmers, who had a vested interest, but also to federal scientists and even to the courts. No one wondered what was happening or why. The only question asked and resolved was whether the regional and national economy could afford the enjoinment of the world's largest copper plant. Ironically, not only were dangers willfully ignored in the name of giant corporate interests, but at one point in time—prior to 1848 and prior to the wholesale conversion of legal precedent in the United States—Anglo-American and British common law had actually offered adequate protection vis-à-vis nuisance injunctions against any of the known dangers. That is, if the history of western mining in the nineteenth century in general, and patterns that emerged from it in Montana specifically, show us anything at all, it is that as a society we probably knew everything we needed to know about protecting people and the environment from the worst impacts of hard-rock mining and the tendencies of a mining society in 1848. Instead, with the lure of a new logic of natural resource exploitation, the economic justifications presented by mass-production ore processing rejected the common knowledge of nuisance injunctions in favor of controlled uncertainty. The twentieth-century US human ecology did not result from ignorance but from willful forgetting. Neither human health nor biological systems have yet recovered.

NOTES

1. Nye, *Electrifying America*.

2. Hoffman, "Butte Ore Lodes," 260–70.

3. Ibid., table on 268–69.

4. Edward Dwyer Peters, *Modern Copper Smelting*, 7th ed. (New York: Scientific Publishing, 1895). See especially chapter 1, "Copper and Its Ores," 1-18.

5. Quivik, "Smoke and Tailings," 164–203.

6. Peters, *Modern Copper Smelting*, see especially chapter 5, "The Preparation of Ores for Roasting," 87–103.

7. Peters, *Modern Copper Smelting*, especially chapters 6, 7, and 8 on roasting ores, 104–223; quotation on 105.

8. Ibid.; Hoffman, "Butte Copper Ores."

9. H. O. Hoffman, "Notes on the Metallurgy of Copper of Montana," *Transactions of the American Institute of Mining Engineers*, 34 (New York: AIME, 1904), 268–69.

10. Summarized from Quivik, "Smoke and Tailings."

11. Calculated using the annual production figure in Stevens and Weed, *Copper Handbook*, 1594; Hoffman, "Butte Copper Ores," 268–69.

12. Ulke, "Characteristic American Metal Mines."

13. "Amended Complaint," 7, filed February 7, 1902, Thomas O. Miles and T. Clowes Miles v. Colorado Smelting & Mining Company (Miles v. Colorado, #195), Civil Case File #195, 1890–1912, US District Court, Butte, Record Group 21, NARA-PR.

14. "Answer," 11–12, filed 1903, Miles v. Colorado, Civil Case File #195, 1890–1912, US District Court, Butte, Record Group 21, NARA-PR. One of the requirements in an equity case such as this one was that any damages had to be specifically determined and associated with the actions of the defendant.

15. Ibid.

16. Miles v. Colorado, 195, Civil Case File #195, 1890–1912, US District Court, Butte, Record Group 21, NARA-PR.

17. "Instructions," Western Loan & Savings Company v. Colorado Smelting & Mining Company, Civil Case File #196, 1890–1912, US District Court, Butte, Record Group 21, NARA-PR.

18. "Complaint," filed September 3, 1903, "Trial Transcript," Hugh Magone v. Colorado Smelting and Mining Company, a corporation, Anaconda Copper Mining Company, a corporation, Colusa-Parrot Mining & Smelting Company, a corporation, Parrot Silver & Copper Company, a corporation, Montana Ore Purchasing Company, a corporation, Butte & Boston Consolidated Mining Company, a corporation (Magone v. Colorado et al., #222), Civil Case File #222, 1890–1912, US District Court, Butte, Record Group 21, NARA-PR; Quivik, "Smoke and Tailings," 299–301.

19. "Complaint," Magone v. Colorado et al., #222 (Heinze quotes); "Separate Answers," and "Brief for the Defendants: Colorado Smelting & Mining Company, Parrot Silver & Copper Company, and Butte and Boston Consolidated Mining Company on Final Hearing," 6, both in Magone v. Colorado et al., Civil Case File #222, 1890–1912, US District Court, Butte, Record Group 21, NARA-PR.

20. On the environmental impacts of all forms of mining in the United States, see Smith, *Mining America*. On the need for water in gold mining, see Rohrbough, *Days of Gold,* 203.

21. See Donald Worster, *River of Empire: Water, Aridity and the Growth of the American West* (New York: Pantheon, 1986), esp. 87–96, 104–11, 160–63. See also Ted Steinberg, *Nature Incorporated: Industrializing the Waters of New England* (New York: Cambridge University Press, 1991).

22. Cited in Gregory Yale, *Legal Titles to Mining Claims and Water Rights in California under the Mining Law of Congress, 1866* (San Francisco: A. Roman, 1867), 141, 145. These

laws were already under assault by industrialists in Massachusetts and up and down the eastern seaboard. See, for example, Steinberg, *Nature Incorporated*.

23. Steinberg, *Nature Incorporated*.

24. Lindley, *Treatise*, 2046–63.

25. Donald J. Pisani, *Water, Land, and Law in the West: The Limits of Public Policy, 1850–
1920* (Lawrence: University Press of Kansas, 1996), esp. chapter 2, "The Origins of Western
Water Law: Case Studies from Two California Mining Districts." See also Donald J. Pisani,
"'I Am Resolved Not to Interfere, but Permit All to Work Freely': The Gold Rush and
American Resource Law," *California History*, special issue on Mining and Economic Devel-
opment in Gold Rush California 77, no. 4 (Winter 1998–99): 123–48, esp. 124–26 (quota-
tion on 124); Elliott West, *The Way to the West: Essays on the Central Plains* (Albuquerque:
University of New Mexico Press, 1995), chapter 4, "Stories," 127–66.

26. Pisani, "I Am Resolved," 136–37.

27. Yale, *Legal Titles*, 138–39.

28. George A. Blanchard and Edward P. Weeks, *The Law of Mines, Minerals, and Mining Water Rights: A Collection of Select and Leading Cases on Mines, Minerals, and Mining
Water Rights, with Notes* (San Francisco: Sumner Whitney, 1877), 727–29; Yale, *Legal Titles*,
136.

29. Blanchard and Weeks, *Law of Mines*, 728–29.

30. Ibid., 729.

31. Yale, *Legal Titles*, 137.

32. Blanchard and Weeks, *Law of Mines*, 696.

33. Charles F. Wilkinson, *Crossing the Next Meridian: Land, Water, and the Future of the
West* (Covelo, CA: Island, 1992).

34. Christine Rosen, "Differing Perceptions of the Value of Pollution Abatement across
Time and Place: Balancing Doctrine in Pollution Nuisance Law, 1840–1906," *Law and
History Review* 11, no. 2 (1993): 303–81; Horwitz, *Transformation of American Law*, 74–78;
Thomas G. Andrews, *Killing for Coal: America's Deadliest Labor War* (Cambridge, MA:
Harvard University Press, 2008).

35. Lindley, *Treatise*, 2049–55.

36. Ibid., 2055–56.

37. Quoted in Smith, *Mining America*, 48.

38. Lindley, *Treatise*, 2057.

39. Ibid., 2058.

40. Ibid.

41. Ibid., 2053–54.

42. Ibid., 2079.

43. Ibid.

44. Ibid.

45. "Complaint," Magone v. Colorado et al., #222, NARA-PR.

46. "Transcript of Testimony Taken before Oliver T. Crane, Esq., Master in Chancery and Examiner," 71–73, Magone v. Colorado et al., Civil Case File #222, 1890–1912, US District Court, Butte, Record Group 21, NARA-PR.

47. Ibid., 110–17.

48. "Findings of Facts Proposed by the Montana Ore Purchasing Company," and "Argument," 6, 9, both in Magone v. Colorado et al., Civil Case File #222, 1890–1912, US District Court, Butte, Record Group 21, NARA-PR.

49. "Findings of Facts Proposed by the Montana Ore Purchasing Company," and "Argument," 6, 9, both in Magone v. Colorado et al., Civil Case File #222, 1890–1912, US District Court, Butte, Record Group 21, NARA-PR.

50. Quoted in Lindley, *Treatise*, 2080–82.

51. Quotations from "Memorandum Order for Decree," 2, filed February 21, 1910, Magone v. Colorado et al., Civil Case File #222, 1890–1912, US District Court, Butte, Record Group 21, NARA-PR; Quivik, "Smoke and Tailings," 305.

52. Donald MacMillan, *Smoke Wars: Anaconda Copper, Montana Air Pollution, and the Courts, 1890–1920* (Helena: Montana Historical Society Press, 2000), 85–89, 101–2; Quivik, "Smoke and Tailings," 308–11; *186 Federal Reporter*, vol. 186, in *Cases Argued and Determined in the Circuit Courts of Appeals and District Courts of the United States, June-July 1911* (St. Paul: West, 1911), 789–827.

53. MacMillan, *Smoke Wars* 102–3; *186 Federal Reporter*, 789–827.

54. MacMillan, *Smoke Wars* 103; Quivik, "Smoke and Tailings," 309; *186 Federal Reporter*, 789–827.

55. MacMillan, *Smoke Wars* 103; Quivik, "Smoke and Tailings," 309; *186 Federal Reporter*, 789–827.

56. *186 Federal Reporter*, 789–827.

57. Ibid.

58. Ibid.

59. Unless, of course, you were the company itself, which as recently as 1903 had closed its entire operation for almost two months to force the state legislature back into session to protect it from the apex litigation brought by Heinze. Ibid.

60. D. E. Salmon, "Arsenical Poisoning from Smelter Smoke in Deer Lodge Valley, Montana," *American Veterinary Review* 39 (April 1911): 14–22, (June 1911): 245–60, (August 1911): 517–38; 40 (November 1911): 164–78, (February 1912): 579–90, (March 1912): 739–47; 41 (May 1912): 164–71, (June 1912): 300–308, (July 1912): 395–421.

> If you can't grow it, you have to mine it.
>
> National Mining Association bumper sticker, 1995

AGAIN AND AGAIN when I first began this project, I encountered decided defensiveness about mining. Those connected to the practice felt a need to remind an increasingly ecologically minded US society that almost everything they do depends on mining. Similarly, when I began to try to make sense of what has happened in Montana and to critique engineers and capitalists alike on mining's wanton impacts, I would be asked if I would prefer to live in a world without electricity. These reactions to my effort to make sense of mining in the US West and to understand it through an environmental lens reflect the basic failure of our historical storytelling about mining, a failure this book makes an effort to address.

The rush to exploit the mineral value embedded in the western mountains of the United States has often been told as a "just-so" story in which western miners pursued a natural wealth that common sense all but demanded they retrieve. Metals rested in places in the landscape, and miners went and got them. For much of mining history, getting the ores was the setting within which other developments—technological, social, economic, and cultural—took place. But as this book has demonstrated, the getting of ores was the plot. The environmental story of mining took place when miners invested in a hoped-for but unknown resource. The formation of the US metal mining industry during the nineteenth century, perhaps among the most important economic developments and environmental commitments in US history, did not take place as the predictable exploitation of a given natural wealth. Instead, from the start the metal mining industry made its own history by gambling on the unknown, gambling on ore, and then adjusting to the consequences or stepping aside for more ambitious developers.[1]

At each step along the way, mining work took place out of sequence with metal production, a curious set of circumstances (perhaps more common than not) where

This 2006 NASA figure of Butte, Montana, shows the massive Berkeley Pit filling with dark, acid-saturated mine water and the oceans of tailings where Butte Hill once stood. From the 1950s into the 1980s, the Anaconda Copper Company worked over every square inch of the hill with mass-destruction techniques, allowing it to extract even trace amounts of copper from huge piles of dirt and rock. Courtesy, NASA.

plans and investments were made on the basis of mineral resources as they existed in the market rather than as they existed in the ground. Mining's environmental story was driven by miners' efforts to close that gap. Gold miners worked against a persistent uncertainty, never satisfied by a single successful washing of creek gold and structurally alienated from the cumulative products of their mining districts and regions. Gold seekers not only mostly failed to benefit materially from their hard labor, but the long-term value of their work went unremunerated and has been historically unrecognized. The mapping of the mineral west by mining rushes during the two decades between 1850 and 1870 represented a singular contribution of knowledge that simply could not have been afforded, never mind replicated, by the hard-rock industry that followed. Indeed, it is fair to say that without this work, the hard-rock industry would never have been launched in the 1860s; and the economic and technological changes that flowed from gold and silver lode mining would have been slowed, if they had emerged at all.

Once mapped, however, the western landscape did not simply yield its contents to lode miners. It certainly whetted their appetites, but the lode miners faced new challenges of creating underground workspaces and managing efficient mills on the surface. Even after the federal government sanctioned lode mining activities with the 1866 mining law, the industry invested much capital and energy but could not seem to produce increasing amounts of gold or silver. The introduction of new techniques and new understandings of minerals' nature by professional mining engineers in the 1870s only partially helped. The federal government was loathe to initiate the state-organized mining efforts the practices brought west by the professional engineers required, but the revised federal mining law did accommodate the larger investments of capital necessary to make lode mining viable under the changing conditions of the 1880s. Railroad subsidies and the ability to move freight helped as well. However, private enterprise on public land represented a Garrett Hardin–like nightmare of shared resource diminution and created new uncertainties and new levels of competition. By the 1880s, all but the wealthiest investors had been priced out of the most productive hard-rocks locations around the West. The nature of hard-rock mining also created patterns of investment and reinvestment in which mining companies regularly expanded the scope and scale of their work in an ongoing effort to catch up with their financial investments before the ores ran out.

When Marcus Daly uncovered rich copper ores in his Anaconda silver mine in the early 1880s, the cycles of iterative change that had marked the silver mining industry accelerated sharply. The combination of larger investments, low-grade ores, and a cheaper metal product led Daly and then several other well-financed mine owners to launch an industry that—for the first time in US history—was capable of competing for the global copper market. These efforts flooded the western world with copper at historic prices and volumes and contributed to the rise and spread of electrification. The need to control costs in the face of uncertainty and expensive competition over ore ownership encouraged the largest companies to integrate both horizontally and vertically in the region. Patterns of modern business arrived early in the copper industry.

Like the gold miners, the early copper industry had no intention of creating this set of circumstances. It was motivated by a need to keep up with its own uncertain and expensive local operations. However, by these events, copper became a commodity of such scale and importance to modern life that financiers on the East Coast and in Europe began to put their resources and influence into the industry. The means by which they organized capital to invest in and expand these expensive undertakings—selling watered-down stock and manipulating copper markets— reflected an expense that seemed to exceed even the wealthiest capitalist's investments.[2] Financiers, too, found mining an uncertain undertaking. Their ownership

"Berkeley Pit & Butte Hill copper mining operation," circa 1955. The company literally excavated the entire mass that was once Butte Hill, truckload by truckload. Extraction has been under way again since the 1990s as pieces of Butte's Boulder Batholith are shipped to Japan and processed for their trace copper. Courtesy, Montana Historical Society, catalog #lot 26 B1 F10 Berkeley Pit, Helena.

of ores was under constant threat in mining fields with more than one mining company. When they worked to consolidate their holdings, the scale and scope of their impacts on the surrounding communities—even in faraway mining-oriented Montana—exceeded the limits of natural ecosystems and social tolerance. Only the emergent industrial-capital–oriented spirit of adjudication saved these companies from impending, and one might say environmentally and socially rational, demise. Because of the mining-oriented water rights and what might be called the commodification of nuisance, copper producers and other miners and processors—and eventually all producers—were given almost free reign to create some of the largest toxic waste dumps in the United States during the twentieth century.

Like the railroads studied by historian Richard White, the mining industry was "built ahead of demand." Rather than view the rise of copper in the late nineteenth century as a tribute to the rationalizing forces of capitalism, the copper industry became a "corporate container for financial manipulation and political network-ing."[3] As a business model, it is fair to say that little was different in mining. Men and women did not invest in metal production, from the gold rush until the Anaconda, with the goal of producing metal for the sake of that metal; rather, they were pro-ducing exchange value every time. Their interest in the metal was not in its use but in its trade. Gold miners sought direct access to wealth buried in the ground under the creek beds, specie to spend on the market which came to them magically wher-ever gold began to be produced. The silver lode miners sought similar ends, a fact made especially obvious by the demise of the silver industry in 1893 when the United States stopped minting silver and put US currency on a gold standard.

Copper producers sought wealth as well, producing more copper than anyone could consume between the late 1870s and the late 1880s but ultimately creating new levels of demand for copper in the process. But just as placer gold miners suf-fered persistent uncertainty at the same time their collective work put a lot of gold into circulation, copper industry giants also suffered persistent uncertainty at the same time their productive activities put unprecedented volumes of copper into the commodities market. Where industry experts saw investments in dead work and risk abatement and hours, if not days, of processing, the average urban consumers believed they saw natural abundance. By 1915, the Anaconda Copper Company was producing popular pamphlets titled "From the Mine to the Consumer," in which it depicted a sanitized and labor-free ore processor environment and, in a gleeful nar-rative, suggested that the entire process rested on the natural wealth buried beneath Butte Hill—a natural wealth that made life in the United States easier.[4] Modern US urbanites became more and more dependent on copper at the same time they were encouraged to know less and less about how it was produced and about the conse-quences of their dependencies.

By the end of the twentieth century, these patterns of dependency on mining had been replicated many times over. It is barely mentioned or remembered, for exam-ple, how much mining and processing were necessary to make possible the engi-neering of an atomic weapon in the 1940s, never mind the massive arms buildup of the Cold War. The processing was, incidentally, dependent on an overproduc-tion of electrical power in the 1930s, as the US government turned hydropower into electrical energy, ideally to bring the essentially urban power system into rural areas. Similarly, it is rarely noted that the rise of commercial air transportation depended on both the availability of cheap electrical energy in the Pacific Northwest and the mining and processing of bauxite to make aluminum. Anaconda's and other copper processors' overproduction of arsenic also contributed to a growing and expanding pesticide industry that, like all of the technologies listed here, would flourish after World War II.

Today, one hardly considers the extensive mining inputs found in our most ubiquitous twenty-first-century technologies; the personal computer and the smart phone consume enormous volumes of cerium and other rare earth metal minerals whose impacts are devastating nations in central Africa, for example. Indeed, years of socializing our use of more and more technologies created a complex and layered masking of the true character of mining and the dependencies and relationships it created. The very ideals that had drawn me to Montana in the early 1990s in the first place, the wilderness and all the trappings of freedom that felt like they should come with it, were made possible by this same mining culture. It is neither an accident nor a coincidence that Yellowstone National Park, Glacier National Park, and four of the largest contiguous designated roadless wilderness areas in the lower forty-eight states are within a 100-mile radius of Butte and Anaconda and the nation's largest Superfund site. They are part of the same cultural landscape. The legacy of mining permeates every corner of our modern lives.[5]

Between 1860 and 1910, the United States became a mining society and extraction became its primary economic mode. In the same way agriculture can be considered foundational in the social and economic structure of life in the United States from the British Colonial era through the mid-nineteenth century, this book maps the path by which mining became an increasingly important way of knowing nature in the United States after 1849 and argues that the United States had become a mining rather than an agrarian culture by the end of the nineteenth century. By "becoming a mining society," I do not mean merely that the United States developed an economy and technological infrastructure dependent on mineral resource exploitation but also that the means by which Americans became miners became an important part of how the culture knows nature and thus a critical piece of who it is as a people. From this conclusion, it is possible to discern that the arsenic in the Clark Fork River was neither an accident nor an oversight but an accurate expression of the emergent forces of economic power to which the nation-state committed itself with all its intentions by the dawn of the twentieth century. It might then be said that the crucial turning point in the body politic of US society, and the origins of today's emergent global society, was not the closing of the agricultural frontier in the 1890s but rather the spawning of the mining frontier in California in the mid-nineteenth century.[6]

NOTES

1. For the economic importance of metal production, see Gavin Wright, "The Origins of American Industrial Success, 1879–1940," *American Economic Review* 80 (September 1990): 651–68.

2. Historian Michael Malone, more than any other writer about Butte, detailed this later tendency in all its ugly and corrupt glory, but Malone's rapt attention to the several individuals involved and the corporate machinations in New York City led him to overlook the systemic character of these outcomes and their origins in the very heart of the mining relationship with nature. See Malone, *Battle for Butte,* esp. chapter 7, "'Consolidation': The Amalgamated and the Independents," and chapter 8, "The Battle for Butte: 1901–1906."

3. Richard White, *Railroaded: The Transcontinentals and the Making of Modern America* (New York: W. W. Norton, 2011), xxvii–xxviii.

4. Anaconda Company, "From the Mine to the Consumer" (Anaconda, MT: Standard Publishing, 1915).

5. http://www.abc.net.au/news/2009-02-17/mobile-phone-metal-funding-dr-congo-conflict/297878; accessed April 12, 2011.

6. Frederick Jackson Turner, "The Significance of the Frontier in American History," in Martin Ridge, ed., *Frederick Jackson Turner: Wisconsin's Historian of the Frontier* (Madison: State Historical Society of Wisconsin, 1986), 41–77; Paul, *Mining Frontier.*

ARCHIVES AND SPECIAL COLLECTIONS

American Heritage Center, Economic Geology Collection,
University of Wyoming Libraries, Laramie
Anaconda Historical Society, Anaconda, MT
Arthur Lakes Library, Colorado School of Mines, Golden
Butte Historical Society, Butte, MT
Henry E. Huntington Library, Art Collections, and Botanical
Gardens, San Marino, CA
James J. Hill Reference Library, St. Paul, MN
Linda Hall Library of Science, Engineering, and Technology,
Kansas City, MO
Minnesota Historical Society, Minneapolis
Montana Historical Society Archives, Helena
National Archives Pacific Alaska Region, Seattle, WA
Special Collection, Maureen and Mike Mansfield Library,
University of Montana, Missoula
Special Collections, New York Public Library, New York, NY
Western History and Genealogy, Denver Public Library,
Denver, CO
William Andrews Clark Memorial Library, UCLA, Los
Angeles, CA

BOOKS, CHAPTERS IN BOOKS, ARTICLES, AND THESES

Alt, David, and Donald W. Hyndman. *Roadside Geology of
Montana*. Missoula: Mountain Press, 1986.
Andrews, Thomas G. *Killing for Coal: America's Deadliest
Labor War*. Cambridge, MA: Harvard University Press, 2008.
Agricola, Georgius. *De Re Metallica*. Translated by Herbert
Clark Hoover and Lou Henry Hoover. New York: Dover
Publications, 1950 [1912].
"The Baltimore Copper Works." *Engineering and Mining
Journal* 30 (1881): 87–88.
Barrows, Willard. "To Idaho and Montana: Wanderings There:
Returning." *Boston Review* 26 (1865): 118–32.
Bartlett, J. C. "American Students of Mining in Germany."
Engineering and Mining Journal 23 (1877): 257–58.

Blanchard, George A., and Edward P. Weeks. *The Law of Mines, Minerals, and Mining Water Rights: A Collection of Select and Leading Cases on Mines, Minerals, and Mining Water Rights, with Notes.* San Francisco: Sumner Whitney, 1877.

Bronin, Andrew. *California Gold Rush, 1849.* New York: Viking, 1972.

Brown, E. G. "The Ore Deposits of Butte City." In *Transactions of the American Institute of Mining Engineers*, 543–58. New York: Scientific Publishing, 1894.

Browne, J. Ross. *Report on the Mineral Resources of the States and Territories West of the Rocky Mountains.* Washington, DC: Government Printing Office, 1868.

Browne, J. Ross, and James W. Taylor. *Reports upon the Mineral Resources of the United States.* Washington, DC: Government Printing Office, 1869.

Brunton, D. W. *Technical Reminiscences.* New York: Mining and Scientific Press, 1915.

Burlingame, Merrill, and Kenneth Ross Toole. *A History of Montana.* New York: Lewis Historical Publishing, 1957.

Chandler, Alfred. *The Visible Hand: The Managerial Revolution in American Business.* Cambridge, MA: Harvard University Press, 1977.

Clark, William Andrews. "Centennial Address on the Origin, Growth and Resources of Montana." Delivered at the Centennial Exposition, Philadelphia, PA, October 11, 1876. In *Contributions to the Historical Society of Montana, with Its Transactions, Act of Incorporation, Constitution, Ordinances, Officers and Members*, 45–60. Helena, MT: State Publishing Company, 1896.

Colby, William E. "The Extralateral Right: Shall It Be Abolished?" *California Law Review* 5, no. 1 (1916): 18. http://dx.doi.org/10.2307/3473991.

Colby, William E. "The Origin and Development of the Extralateral Right in the United States." *California Law Review* 4, no. 6 (1916): 437–64. http://dx.doi.org/10.2307/3473690.

Connolly, C. P. "The Story of Montana." *McClure's Magazine* 28 (1906): 346–61.Connolly, C. P. "The Story of Montana II: The Treasure of Butte Hill and Development of the Great Copper Industry—Beginnings of the Clark-Daly Feud." *McClure's Magazine* 27, no. 5 (1906): 451–65.

Connolly, C. P. "The Story of Montana III: A Summary of Important Events in Montana Politics from 1894–1896." *McClure's Magazine* 27, no. 6 (1906): 629–39.

Contributions to the Historical Society of Montana; with Its Transactions, Act of Incorporation, Constitution, Ordinances, Officers and Members, vol. I. Helena, MT: Rocky Mountain Publishing, 1876.

"The Copper Mines of Butte, Montana." *Engineering and Mining Journal* 23 (1877): 272–73.

Coxe, Eckley B. "Mining Engineering." *Engineering and Mining Journal* 26 (1878): 247–49.

Coxe, Eckley B. "Secondary Technical Education." *Engineering and Mining Journal* 27 (1879): 144–46.

Cronon, William. *Nature's Metropolis: Chicago and the Great West.* New York: W. W. Norton, 1991.

Cronon, William, ed. *Uncommon Ground: Rethinking the Human Place in Nature.* New York: W. W. Norton, 1996.

Cronon, William, George Miles, and Jay Gitlin, eds. *Under an Open Sky: Rethinking America's Western Past.* New York: W. W. Norton, 1992.

Curtis, Kent. "Producing a Gold Rush: National Ambitions and the Northern Rocky Mountains, 1853–1863." *Western Historical Quarterly* 40, no. 3 (2009): 275–97.

Davis, Watson. *The Story of Copper.* New York: Century, 1924.

Dimsdale, Thomas J. *The Vigilantes of Montana: A Correct History of the Chase, Capture, Trial and Execution of Henry Plummer's Notorious Road Agent Band.* Butte, MT: W. F. Bartlett, 1915 [1865].

Donahue, Brian. *The Great Meadow: Farmers and the Land in Colonial Concord.* New Haven, CT: Yale University Press, 2004.

Douglas, James. *Cantor Lectures on Recent American Methods and Appliances Employed in the Metallurgy of Copper, Lead, Gold, and Silver.* London: William Trounce, 1895.

Douglas, James. "The Copper Industry of the United States." *Iron Age* (January 2, 1896): 3–6.

Douglas, James. *The Copper-Resources of the United States. Read Provisionally at the New York Meeting, September 1890, and Completed with the Statistics for the Current Year.* Author's ed., 1891.

Douglas, James. "Letters from the West: IX." *Engineering and Mining Journal* 33 (1882): 219.

Dunne, E. F. "The United States Mining Law." *Engineering and Mining Journal* 9 (1870): 2–4.

Edgar, Henry. "The Journal of Henry Edgar." In *Contributions to the Historical Society of Montana; with Its Transactions, Officers and Members,* vol. 3. Helena, MT: State Publishing Company, 1900.

Eifler, Mark A. *Gold Rush Capitalists: Greed and Growth in Sacramento.* Albuquerque: University of New Mexico Press, 2002.

Eilers, Anton. "Silver Smelting in Montana: General Considerations." *Engineering and Mining Journal* 12 (1871): 241–42.

Eilers, Anton. "Silver Smelting in Montana, II: The Works of Argenta." *Engineering and Mining Journal* 12 (1871): 257–58.

Eilers, Anton. "Western Montana, Deer Lodge City, August 6, 1871." *Engineering and Mining Journal* 12 (1871): 168, 177–78, 191.

Elliot, Russell R. *Servant of Power: A Political Biography of William M. Stewart.* Reno: University of Nevada Press, 1983.

Emmons, David M. *The Butte Irish: Class and Ethnicity in an American Mining Town, 1875–1925.* Urbana: University of Illinois Press, 1989.

Emmons, S. F. *Notes on the Geology of Butte, Montana.* New York: American Institute of Mining Engineers, 1887.

Ewing, R. C. "Report of Hon. R.C. Ewing, Helena, Montana Territory, October 12, 1865." Pamphlet Collection 3893. Helena: Montana Historical Society, 1865.

Fell, James. *Ores to Metals: The Rocky Mountain Smelting Industry.* Lincoln: University of Nebraska Press, 1979.

Fisk, James L. *Expedition of Captain Fisk to the Rocky Mountains: Letter from the Secretary of War in Answer to a Resolution of the House of February 26, Transmitting Report of Captain Fisk of His Late Expedition to the Rocky Mountains and Idaho.* 36th Congress, 1st Session. Washington, DC: Government Printing Office, 1864.

Garnett, R. H. T., and N. C. Bassett. "Placer Deposits." In *Economic Geology One Hundredth Anniversary Volume: 1905–2005: Society of Economic Geologists,* ed. Jeffrey W. Hedenquist, John F.H. Thompson, Richard J. Goldfarb, and Jeremy P. Richards, 813–44. Littleton, CO: Society of Economic Geologists, 2005.

Glasscock, Carl B. *The War of the Copper Kings: Builders of Butte and Wolves of Wall Street.* New York: Grosset and Dunlap, 1935.

Goldfarb, Richard J., T. Baker, B. Dubé, D. I. Groves, C. J. R. Hart, F. Robert, and P. Gosselin. "World Distribution, Productivity, Character, and Genesis of Gold Deposits in Metamorphic Terranes." In *Economic Geology One Hundredth Anniversary Volume: 1905–2005: Society of Economic Geologists,* ed. Jeffrey W. Hedenquist, John F.H. Thompson, Richard J. Goldfarb, and Jeremy P. Richards, 407–50. Littleton, CO: Society of Economic Geologists, 2005.

Goodman, David. *Gold Seeking: Victoria and California in the 1850s.* Stanford, CA: Stanford University Press, 1994.

Gordon, Robert B., and Patrick M. Malone. *The Texture of Industry: An Archeological View of the Industrialization of North America.* New York: Oxford, 1994.

Greever, William S. *The Bonanza West: The Story of the Western Mining Rushes, 1848–1900.* Norman: University of Oklahoma Press, 1963.

Groves, David I., Kent C. Condie, Richard J. Goldfarb, Jonathan M.A. Hronsky, and Richard Vielreicher. "Secular Changes in Global Tectonic Processes and Their Influence on the Temporal Distribution of Gold-Bearing Mineral Deposits." *Economic Geology and the Bulletin of the Society of Economic Geologists* 100, no. 2 (March 2005): 203–24. http://dx.doi.org/10.2113/gsecongeo.100.2.203.

Hague, James D., with geological contributions by Clarence King. *Mining Industry,* vol. 3 of the United States Geological Exploration of the Fortieth Parallel. Washington, DC: Government Printing Office, 1870.

Hamilton, James McClellan. *From Wilderness to Statehood: A History of Montana, 1805–1900.* Portland, OR: Binfords and Mort, 1957.

Hays, Samuel P. *Beauty, Health, and Permanence: Environmental Politics in the United States, 1955–1985*. New York: Cambridge University Press, 1987. http://dx.doi.org/10.1017/CBO9780511664106

Hewitt, A. S. "American Institute of Mining Engineers." *Engineering and Mining Journal* 13 (1872): 338–39.

Hittell, John S. *Mining in the Pacific States of North America*. San Francisco: H. H. Bancroft, 1861.

Hofman, H. O. "Notes on the Metallurgy of Copper of Montana." *Transactions of the American Institute of Mining Engineers* 34 (1904): 258–316.

Holliday, J. S., and William Swain. "The World Rushed." In *The California Gold Rush Experience*. New York: Simon and Schuster, 1981.

Horwitz, Morton. *The Transformation of American Law, 1790–1860*. Cambridge, MA: Harvard University Press, 1977.

Hughes, Thomas P. *Networks of Power: Electrification in Western Society, 1880–1930*. Baltimore: Johns Hopkins University Press, 1983.

Hurtado, Albert L. *Indian Survival on the California Frontier*. New Haven, CT: Yale University Press, 1988.

Hyde, Charles K. *Copper for America: The United States Copper Industry from Colonial Times to the 1990s*. Tucson: University of Arizona Press, 1998.

Isenberg, Andrew. *Mining California: An Ecological History*. New York: Hill and Wang, 2005.

Israel, Paul. *Edison: A Life of Invention*. New York: John Wiley and Sons, 2000.

Jackson, Donald Dale. *Gold Dust: The California Gold Rush and the Forty-Niners*. Boston: Allen and Unwin, 1980.

John, David A., Albert H. Hofstra, and Ted G. Theodore. "Part 1: Regional Studies and Epithermal Deposits." *Economic Geology: A Special Issue Devoted to Gold in Northern Nevada* 98, no. 2 (April 2003): 225–34. http://dx.doi.org/10.2113/gsecongeo.98.2.225.

Johnson, Steven. *Emergence: The Connected Lives of Ants, Brains, Cities, and Software*. New York: Scribner, 2001.

Johnson, Susan Lee. *Roaring Camp: The Social World of the California Gold Rush*. New York: W. W. Norton, 2000.

Keyes, William W. "The Defects of the Mining Law." In *Statistics of Mines and Mining in the States and Territories West of the Rocky Mountains, 1873*, ed. Rossiter W. Raymond, 512–19. Washington, DC: Government Printing Office, 1874.

King, Clarence. *Mining Industry*. United States Geological Exploration of the Fortieth Parallel, vol. 3. Washington, DC: Government Printing Office, 1870.

Langford, Nathaniel Pitt. *Address Delivered before the Grand Lodge of Montana at Its Third Annual Communication, in the City of Virginia, October 8, A.D. 1867*. Helena, MT: "Herald" Book Establishment, 1868.

Langford, Nathaniel Pitt. *Vigilante Days and Ways: The Pioneers of the Rockies, the Makers and Making of Montana, Idaho, Oregon, Washington, and Wyoming*, vol. 1. New York: D. D. Merrill, 1893.

Lapp, Rudolph M. *Blacks in Gold Rush California*. New Haven, CT: Yale University Press, 1977.

LeCain, Timothy J. *Mass Destruction: The Men and Giant Mines That Wired America and Scarred the Planet*. New Brunswick, NJ: Rutgers University Press, 2009.

Lefebvre, Henri. *The Production of Space*. Trans. Donald Nicholson-Smith. Cambridge, MA: Blackwell, 1991 [1974].

Leifchild, John R. *Cornwall: Its Mines and Miners with Sketches of Scenery Designed as a Popular Introduction to Metallic Mines*. London: Longman, Brown, Green, and Longmans, 1855.

Levinson, Robert E. *The Jews in the California Gold Rush*. New York: Ktav, 1978.

Levy, JoAnn. *They Saw the Elephant: Women and the California Gold Rush*. Norman: University of Oklahoma Press, 1992.

Lindley, Curtis H. *A Treatise on the American Law Relating to Mines and Mineral Lands within the Public Land States and Territories and Governing the Acquisition and Enjoyment of Mining Rights in Lands of the Public Domain*, vol. I. San Francisco: Bancroft-Whitney, 1897; 3rd ed., San Francisco: Bancroft-Whitney, 1914.

Lyman, Benjamin Smith. "On the Importance of Surveying in Geology." In *Transactions of the American Institute of Mining Engineers*, vol. 1: *May 1871–February 1873*, ed. Rossiter W. Raymond, 183–92. Philadelphia: AIME, 1873.

MacMillan, Donald. *Smoke Wars: Anaconda Copper, Montana Air Pollution, and the Courts, 1890–1920*. Helena: Montana Historical Society Press, 2000.

Madsen, Brigham D. *The Shoshoni Frontier and the Bear River Massacre*. Salt Lake City: University of Utah Press, 1985.

Malone, Michael P. *The Battle for Butte: Mining and Politics on the Northern Frontier, 1864–1906*. Helena: Montana Historical Society Press, 1995.

Malone, Michael P. *James J. Hill: Empire Builder of the Northwest*. Norman: University of Oklahoma Press, 1996.

Malone, Michael P. "Midas of the West: The Incredible Career of William Andrews Clark." *Montana: Magazine of Western History* 33, no. 4 (1983): 2–17.

Malone, Michael P., and Richard B. Roeder. *Montana: A History of Two Centuries*. Seattle: University of Washington Press, 1976.

Mann, Ralph. *After the Gold Rush: Society in Grass Valley and Nevada City California, 1849–1870*. Stanford, CA: Stanford University Press, 1982.

Marcossen, Isaac F. *Anaconda*. New York: Dodd, Mead, 1957.

Marx, Karl. *Value, Price and Profit*. New York: International Company, 1969.

McNeer, May. *The California Gold Rush*. New York: Random House, 1994.

Mellinger, Philip J. *Race and Labor in Western Copper: The Fight for Equality, 1896–1918*. Tucson: University of Arizona Press, 1995.

Meredith, Emily R. "Bannack and Gallatin City in 1862–1863." In *Historical Reprints: Sources of Northwest History* no. 24, ed. Paul C. Philips. Missoula: Montana State University, 1933.

Michno, Gregory. *Encyclopedia of Indian Wars: Western Battles and Skirmishes, 1850–1890*. Missoula, MT: Mountain Press, 2003.

"Montana." *Engineering and Mining Journal* 8 (1869): 261.

Morley, Virginia F. *James Henry Morley, 1824–1889: A Memorial*. Cambridge, MA: Riverside, 1891.

Morse, Kathryn. *The Nature of Gold: An Environmental History of the Klondike Gold Rush*. Seattle: University of Washington Press, 2003.

Mullan, Captain John. *Report on the Construction of a Military Road from Walla-Walla to Fort Benton*. Washington, DC: Government Printing Office, 1863.

Murphy, Mary. *Mining Cultures: Men, Women, and Leisure in Butte 1914–1941*. Urbana: University of Illinois Press, 1997.

Nash, Roderick. *Wilderness and the American Mind*. New Haven, CT: Yale University Press, 1982.

Newton, Joseph. *Metallurgy of Copper*. New York: John Wiley and Sons, 1942.

Nitze, Henry B.C. "A Description of the South Welsh Method of Copper Smelting as Practiced by the Baltimore Smelting and Rolling Co., Baltimore, Md." BS thesis, Lehigh University, Bethlehem, PA, 1887.

Nye, David E. *Electrifying America: Social Meaning of a New Technology*. Cambridge, MA: MIT Press, 1997.

Nye, David E. *Narratives and Spaces: Technology and the Construction of American Culture*. New York: Columbia University Press, 1997.

O'Farrell, Pete A. *Butte: Its Copper Mines and Copper Kings*. New York: James A. Rogers, 1899.

Ott, Adolph. "On Mining." Translated from advance sheets of the *German-American Encyclopedia. Engineering and Mining Journal* 9 (1870): 233.

Paul, Rodman Wilson. *California Gold*. Lincoln: University of Nebraska Press, 1965.

Paul, Rodman Wilson. *Mining Frontiers of the Far West, 1848–1880*. Albuquerque: University of New Mexico Press, 1974.

Peters, Edward Dyer, Jr. "The Mines and Reduction Works of Butte City, Montana." In *Mineral Resources of the United States, Calendar Years 1883 and 1884*, ed. Albert Williams Jr., 374–98. Report for the Department of the Interior. Washington, DC: Government Printing Office, 1885.

Peters, Edward Dyer, Jr. *Modern Copper Smelting*, 7th ed. New York: Scientific Publishing, 1895.

Peterson, Richard. *The Bonanza Kings: The Social Origins and Business Behavior of Western Mining Entrepreneurs, 1870–1900*. Lincoln: University of Nebraska Press, 1977.

Phillips, Paul C. "Granville Stewart, 'Montana as It Is.' " In *Sources of Northwest History* no. 16, gen. ed. Paul C. Phillips. Missoula: Montana State University, 1931.

Phillips, Paul C., ed. *Forty Years on the Frontier, as Seen in the Journals and Reminiscences of Granville Stuart, Gold-Miner, Trader, Merchant, Rancher and Politician*, vol. 1. Lincoln: University of Nebraska Press, 1977.

Pisani, Donald J. " 'I Am Resolved Not to Interfere, but Permit All to Work Freely': The Gold Rush and American Resource Law." *California History*, special issue on Mining and Economic Development in Gold Rush California 77, no. 4 (1998-99): 123–48.

Pisani, Donald J. *Water, Land, and Law in the West: The Limits of Public Policy, 1850–1920*. Lawrence: University Press of Kansas, 1996.

Pokrovski, Gleb S., and Leonid S. Dubrovinsky. "The S^3 Ion Is Stable in Geological Fluids at Elevated Temperatures and Pressures." *Science* 331, no. 6020 (February 25, 2011): 1052–54. http://dx.doi.org/10.1126/science.1199911. Medline: 21350173.

Quivik, Fredric L. "Smoke and Tailings: An Environmental History of Copper Smelting Technologies in Montana, 1880–1930." PhD diss., University of Pennsylvania, Philadelphia, 1998.

Raymer, Robert George. "A History of Copper Mining in Montana." PhD diss., Northwestern University, Evanston, IL, 1930.

Raymond, Rossiter W. *Biographical Note of James Duncan Hague: A Paper Read before the American Institute of Mining Engineers, at the Chattanooga Meeting, October, 1908*. Chattanooga: Author's ed., 1909.

Raymond, Rossiter W. "Comparison of Mining Conditions To-day with Those of 1872, in Their Relation to Federal Mineral Lands." In *Transactions of the American Institute of Mining Engineers*, vol. 48: *February 1914*, 299–306. New York: AIME, 1915.

Raymond, Rossiter W. "Editorial Correspondence: The Dormant Resources of Montana; Virginia City, Montana Territory, July 29, 1871." *Engineering and Mining Journal* 12 (1871): 113–14.

Raymond, Rossiter W. "Editorial Correspondence: Montana and the Railroad; August 2, 1871." *Engineering and Mining Journal* 12 (1871): 153–54.

Raymond, Rossiter W. "The Geological Distribution of Mining Districts in the United States." In *Transactions of the American Institute of Mining Engineers*, vol. 1: *May 1871– February 1873*, 34–39. Philadelphia: AIME, 1873.

Raymond, Rossiter W. *A Glossary of Mining and Metallurgical Terms*. Easton, PA: American Institute of Mining Engineers, 1881.

Raymond, Rossiter W. "Letters." *Engineering and Mining Journal* 17 (1874): 158.

Raymond, Rossiter W. *Mineral Resources of the States and Territories West of the Rocky Mountains*. Washington, DC: Government Printing Office, 1869.

Raymond, Rossiter W. *The Mines of the West: A Report to the Secretary of the Treasury*. New York: J. B. Ford, 1869.

Raymond, Rossiter W. "Mining Statistics." *Engineering and Mining Journal* 9 (1870): 185.

Raymond, Rossiter W. "The National School of Mines." *Engineering and Mining Journal* 9 (1870): 184.

Raymond, Rossiter W. "The New Mining Law." *Engineering and Mining Journal* 11 (1871): 121.

Raymond, Rossiter W. *Silver and Gold: The Mining and Metallurgical Industries of the United States, with Reference Chiefly to the Precious Metals*. New York: J. B. Ford, 1873.

Raymond, Rossiter W. *Statistics of Mines and Mining in the States and Territories West of the Rocky Mountains, 1869*. Washington, DC: Government Printing Office, 1870.

Raymond, Rossiter W. *Statistics of Mines and Mining in the States and Territories West of the Rocky Mountains, for the Year 1870*. Washington, DC: Government Printing Office, 1872.

Raymond, Rossiter W. *Statistics of Mines and Mining in the States and Territories West of the Rocky Mountains, for the Year 1871*. Washington, DC: Government Printing Office, 1873.

Raymond, Rossiter W. *Statistics of Mines and Mining in the States and Territories West of the Rocky Mountains, 1873*. Washington, DC: Government Printing Office, 1874.

Raymond, Rossiter W. "Time and Man—a Mining Partnership." *Engineering and Mining Journal* 9 (1870): 313.

Reid, Bernard J., and Mary McDougall Gordon. *Overland to California with the Pioneer Line: The Gold Rush Diary of Bernard J. Reid*. Stanford, CA: Stanford University Press, 1983.

Richardson, A. D. "Our New States and Territories: Montana and Idaho." *Beadle's Monthly: A Magazine of To-day* 2 (1866): 277–90.

Rickard, Thomas A. *A History of American Mining*. New York: Macmillan, 1932.

Robbins, William. *Colony and Empire: The Capitalist Transformation of the American West*. Lawrence: University Press of Kansas, 1994.

Robertson, Kenneth, M.E. "Blast-Furnace Slags." In *Transactions of the American Institute of Mining Engineers*, vol.1: *May 1871–February 1873*, 145–60. Philadelphia: AIME, 1873.

Rohrbough, Malcolm J. *Days of Gold: The California Gold Rush and the American Nation*. Berkeley: University of California Press, 1997.

Rosen, Christine. "Differing Perceptions of the Value of Pollution Abatement across Time and Place: Balancing Doctrine in Pollution Nuisance Law, 1840–1906." *Law and History Review* 11, no. 2 (1993): 303–81. http://dx.doi.org/10.2307/743617.

Rosen, Christine Meisner, and Christopher C. Sellers. "The Nature of the Firm: Towards an Ecocultural History of Business." *Business History Review* 73, no. 4 (1999): 577–600.

Rosner, David, and Gerald E. Markowitz. *Deadly Dust: Silicosis and the Politics of Occupational Disease in Twentieth Century America*. Princeton, NJ: Princeton University Press, 1991.

Rothwell, Richard P., M.E. "Remarks on the Waste in Coal Mining." In *Transactions of the American Institute of Mining Engineers*, ed. Rossiter W. Raymond, vol. 1: *May 1871–February 1873*, 45–71. Philadelphia: AIME, 1873.

Rowe, John. *The Hard Rock Men: Cornish Immigrants and the North American Mining Frontier*. New York: Barnes and Noble Books, 1974.

Safford, Jeffrey J. *The Mechanics of Optimism: Mining Companies, Technology, and the Hot Spring Gold Rush, Montana Territory, 1864–1868*. Boulder: University Press of Colorado, 2004.

Sales, Reno H. *Underground Warfare at Butte*. Caldwell, ID: Caxton, 1964.

Scott, James C. *Seeing Like a State: How Certain Schemes to Improve the Human Condition Have Failed*. New Haven, CT: Yale University Press, 1998.

Sears, Marian V. *Mining Stock Exchanges, 1860–1930*. Missoula: University of Montana Press, 1973.

Senkewicz, Robert M. *Vigilantes in Gold Rush California*. Stanford, CA: Stanford University Press, 1985.

Shinn, Charles Howard. *Mining Camps: A Study in American Frontier Government*. New York: Alfred A. Knopf, 1948 [1884].

Shoebotham, H. Minar. *Anaconda: Life of Marcus Daly the Copper King*. Harrisburg, PA: Stackpole, 1956.

Smith, Duane A. *Mining America: The Industry and the Environment, 1800–1980*. Niwot: University Press of Colorado, 1993 [1987].

So, Chil-Sup, and Kevin L. Shelton. "Stable Isotope and Fluid Inclusion Studies of Gold- and Silver-Bearing Hydrothermal Vein Deposits, Cheonan-Cheongyang-Nonsan Mining District, Republic of Korea; Cheonan Area." *Economic Geology and the Bulletin of the Society of Economic Geologists* 82, no. 4 (July 1, 1987): 987–1000. http://dx.doi.org/10.2113/gsecongeo.82.4.987.

Spence, Clark C. *Mining Engineers and the American West: The Lace-Boot Brigade*. New Haven, CT: Yale University Press, 1970.

Steinberg, Ted. *Down to Earth: Nature's Role in American History*. New York: Oxford University Press, 2008.

Steinberg, Ted. *Nature Incorporated: Industrializing the Waters of New England*. New York: Cambridge University Press, 1991.

Stevens, Horace J., and Walter Harvey Weed. *The Copper Handbook: A Manual of the Copper Industry of the World*, vol. 9 for the Year 1909. Chicago: M. A. Donohue, 1910.

Stewart, John. *Thomas F. Walsh: Progressive Businessman and Colorado Tycoon*. Boulder: University Press of Colorado, 2007.

Stine, Jeffrey K., and Joel A. Tarr. "At the Intersection of Histories: Technology and the Environment." *Technology and Culture* 39, no. 4 (1998): 601–40.

Stout, Tom, ed. *Montana: Its Story and Biography*. Chicago: American Historical Society, 1921.

Stuart, Granville. "Montana as It Is." *Sources of Northwest History* no. 16, gen. ed. Paul C. Philips. Missoula: Montana State University, 1931.

Stuart, James, with notes by Samuel T. Hauser and Granville Stuart. "The Yellowstone Expedition of 1863." In *Contributions to the Historical Society of Montana; with Its Transactions, Acts of Incorporation, Constitution, Ordinances, Officers and Members*, vol. 1. Helena, MT: Rocky Mountain Publishing, 1876.

Sutter, Paul. *Driven Wild: How the Fight against the Automobile Launched the Modern Wilderness Movement*. Seattle: University of Washington Press, 2002.

Swenson, Robert W. "Legal Aspects of Mineral Resources Exploitation." In *History of Public Land Law Development*, ed. Paul Gates, 701–28. Public Land Law Review Commission. Washington, DC: US Government Printing Office, 1968.

Tanzer, Michael. *The Race for Resources: Continuing Struggles over Mineral and Fuels*. New York: Monthly Review Press, 1980.

Toole, K. Ross. "A History of the Anaconda Copper Company: A Study in the Relationship between a State, Its People, and a Corporation." PhD diss., UCLA, Los Angeles, 1954.

Turner, Frederick Jackson. "The Significance of the Frontier in American History." In *Frederick Jackson Turner: Wisconsin's Historian of the Frontier*, ed. Martin Ridge. Madison: State Historical Society of Wisconsin, 1986.

Ulke, Titus. "Characteristic American Metal Mines: The Anaconda Copper Mines and Works." *Engineering Magazine* (July 1897): 512–34.

Vinton, Prof. F. H. "Mining Engineering." *Engineering and Mining Journal* 17 (1874): 138–39, 163, 199.

Waltner-Townes, David, James J. Kay, and Nina-Marie Lister. *The Ecosystem Approach: Complexity, Uncertainty, and Managing for Sustainability*. New York: Columbia University Press, 2008.

West, Elliott. *The Contested Plains: Indians, Goldseekers, and the Rush to Colorado*. Lawrence: University Press of Kansas, 1998.

West, Elliott. *The Way to the West: Essays on the Central Plains*. Albuquerque: University of New Mexico Press, 1995.

White, Helen McCann, ed. *Ho! For the Gold Fields: Northern Overland Wagon Trains of the 1860s*. St. Paul: Minnesota Historical Society, 1966.

White, Richard. *"It's Your Misfortune and None of My Own": A New History of the American West*. Norman: University of Oklahoma Press, 1991.

White, Richard. *The Organic Machine: Remaking Nature on the Columbia River*. New York: Hill and Wang, 1996.

White, Richard. *Railroaded: The Transcontinentals and the Making of Modern America*. New York: W. W. Norton, 2011.

Wilkinson, Charles F. *Crossing the Next Meridian: Land, Water, and the Future of the West.* Covelo, CA: Island, 1992.

Williams, Mary Floyd. *History of the San Francisco Committee of Vigilance of 1851: A Study of Social Control on the California Frontier in the Days of the Gold Rush.* Berkeley: University of California Press, 1921.

Winner, Langdon. *The Whale and the Reactor: A Search for Limits in an Age of High Technology.* Chicago: University of Chicago Press, 1986.

Worster, Donald. *Dust Bowl: The Southern Plains in the 1930s.* New York: Oxford University Press, 1979.

Worster, Donald. *Nature's Economy: A History of Ecological Ideas.* New York: Cambridge University Press, 1994.

Worster, Donald. *River of Empire: Water, Aridity and the Growth of the American West.* New York: Pantheon, 1986.

Worster, Donald, et al. "Environmental History: A Roundtable." *Journal of American History* 76, no. 4 (March 1990): 1087–1147. http://dx.doi.org/10.2307/2936586.

Wright, Gavin. "The Origins of American Industrial Success, 1879–1940." *American Economic Review* 80 (September 1990): 651–68.

Wright, James E. *The Galena Lead District: Federal Policy and Practice, 1824–1847.* Madison: Department of History, University of Wisconsin, 1966.

Yale, Gregory. *Legal Titles to Mining Claims and Water Rights in California under the Mining Law of Congress, 1866.* San Francisco: A. Roman, 1867.

Young, Otis E. *Western Mining: An Informal Account of Precious-Metals Prospecting, Placering, Lode Mining, and Milling on the American Frontier from Spanish Times to 1893.* Norman: University of Oklahoma Press, 1970.

Zeihen, Lester G., Richard B. Berg, and Henry G. McClernan. "Geology of the Butte District, Montana." In *Geological Society of America's Centennial Field Guide*, ed. Donald L. Biggs, 57–61. Boulder: Geological Society of America, 1987.

Index